# Human Communication and the Brain

# Human Communication and the Brain

Donald B. Egolf

LEXINGTON BOOKS
*Lanham • Boulder • New York • Toronto • Plymouth, UK*

Published by Lexington Books
A wholly owned subsidiary of The Rowman & Littlefield Publishing Group, Inc.
4501 Forbes Boulevard, Suite 200, Lanham, Maryland 20706
http://www.lexingtonbooks.com

Estover Road, Plymouth PL6 7PY, United Kingdom



British Library Cataloguing in Publication Information Available

**Library of Congress Cataloging-in-Publication Data**

Library of Congress Cataloging-in-Publication Data Available
ISBN 978-0-7391-3963-9 (cloth : alk. paper) — ISBN 978-0-7391-3964-6 (pbk : alk. paper) — ISBN 978-0-7391-3965-3

™ The paper used in this publication meets the minimum requirements of American National Standard for Information Sciences—Permanence of Paper for Printed Library Materials, ANSI/NISO Z39.48-1992.

Printed in the United States of America

# *Table of Contents*

# Figures

# PREFACE

The world communicates to the brain and the brain is affected by the world's messages. The brain in turn communicates to the body and the body affects the world. Just as it was often said that all roads lead to Rome, all messages from the world go to the brain and all responses to those messages emanate from the brain. What now fascinates neuroscientists, scholars from other disciplines, and to an extent, society at large is the communication that occurs within the brain. What happens in the brain when a person sees a smile on another person's face, when a person decides to eat or not eat that piece of chocolate cake, when one is trying to decide to buy the red one or the green one, and when one is deciding to vote for Candidate A or Candidate B, for example? And then there are the cases where free will seems not to exist, when a person is driven as if controlled by uncontrollable forces. There are reflective as well as reflexive messages that the brain produces. Of equal importance is to study what happens in the brain when messages from the outside world are directed at the brain for the purpose of persuasion and control. What happens in the brain when a person begins to believe or is converted to some cause? These are the questions on the agendas of neuroscientists and like-minded scholars from other disciplines.

The purpose of this book is to look at the search for answers to the many issues posed by brain research today. The examination will be from the perspective of a communication scholar, a perspective somewhat specific but hopefully broader than a description of an elephant based on the contours of one leg.

The author of this book gratefully acknowledges the assistance of Sondra Chester and Lynn Cooper for their editing help and Randal McKenzie for doing the medical illustrations. The editorial help of Lenore Lautigar of Lexington Books is also acknowledged. Thanks are also extended to the many people with whom the author has conversed over the years that aroused and maintained his interest in one of the most interesting structures in the world: the human brain.

# *Part I*
# Introduction

The purpose of the chapters in this introduction is to prepare you for reading and comprehending the subsequent chapters. The first chapter documents the exponential increase in the interest in the brain, manifested not only in the research activities of professional and academic scholars, but in the lay world as well. Truly it is the age of the brain.

The second chapter in this introduction seeks to provide a basic understanding of the brain. Included will be notes on the brain's complexity, biographical notes on a number of prominent brain researchers arranged in chronological order, and some elemental physiological concepts will be presented beginning with the neuron as the basic unit of analysis. Finally, some basic neuroanatomical structures will be identified and discussed.

Chapter 3 deals with methodologies that have been developed and deployed in an effort to better understand the brain. It will be shown that developments in neuroscience have paralleled methodological developments. Before the development of modern brain scanning methods, for example, information about the brain was gathered primarily from trauma cases whose post-trauma symptoms were then compared with pre-morbid behaviors, if the patient survived the trauma, or with the post-mortem findings recorded in autopsies.

The purpose of chapter 4 is to discuss the complexity of human communication. This is important because the purpose of this book is to show the relevance of neuroscience research to human communication. At first glance allocating a chapter to human communication may seem excessive. Why all the fuss? But we shall see that a great deal of attention has been paid to this issue. It all begins with defining the concept. And here it may be important to keep in mind the words of Will Durant (1974) who said that to get involved in issues of definition is to "let loose the dogs of philosophical war. For nothing is so difficult as definition, nor anything so severe a test and exercise of mental clarity and skill."

# *Chapter 1*
# The Age of the Brain

Interest in the brain has snowballed in the last two decades. And, like the proverbial snowball metaphor, the interest in the brain has included both acceleration and growth. Five major factors have contributed to this snowballing; they are the Decade of the Brain Project, the Dana Alliance, the Human Brain Project, the Alzheimer's Disease Neuroimaging Initiative Project, and the technological advances in brain imaging. These factors have increased the number of disciplines involved in brain research, have increased dramatically the number of research projects being proposed and conducted, and have led to a "spillover" of interest into the public domain. A brief description of each of the five factors will be given below, followed by a summary of the effects of the five factors.

## The Decade of the Brain Project

A good description of the Decade of the Brain Project can be found in the presidential proclamation declaring the 1990s as the decade of the brain:

> Presidential Proclamation 6158
> July 17, 1990
> By the President of the United States of America
> A Proclamation
>
> The human brain, a 3–pound mass of interwoven nerve cells that controls our activity, is one of the most magnificent—and most mysterious—wonders of creation. The seat of human intelligence, interpreter of senses, and controller of movement, this incredible organ continues to intrigue scientists and laymen alike.

Over the years, our understanding of the brain—how it works, what goes wrong when it is injured or diseased—has increased dramatically. However, we still have much more to learn. The need for continued study of the brain is compelling: millions of Americans are affected each year by disorders of the brain ranging from neurogenetic diseases to degenerative disorders such as Alzheimer's, as well as stroke, schizophrenia, autism, and impairments of speech, language, and hearing.

Today, these individuals and their families are justifiably hopeful, for a new era of discovery is dawning in brain research. Powerful microscopes, major strides in the study of genetics, and advances in brain imaging devices are giving physicians and scientists ever greater insight into the brain. Neuroscientists are mapping the brain's biochemical circuitry, which may help produce more effective drugs for alleviating the suffering of those who have Alzheimer's or Parkinson's disease. By studying how the brain's cells and chemicals develop, interact, and communicate with the rest of the body, investigators are also developing improved treatments for people incapacitated by spinal cord injuries, depressive disorders, and epileptic seizures. Breakthroughs in molecular genetics show great promise of yielding methods to treat and prevent Huntington's disease, the muscular dystrophies, and other life-threatening disorders.

Research may also prove valuable in our war on drugs, as studies provide greater insight into how people become addicted to drugs and how drugs affect the brain. These studies may also help produce effective treatments for chemical dependency and help us to understand and prevent the harm done to the preborn children of pregnant women who abuse drugs and alcohol. Because there is a connection between the body's nervous and immune systems, studies of the brain may also help enhance our understanding of Acquired Immune Deficiency Syndrome.

Many studies regarding the human brain have been planned and conducted by scientists at the National Institutes of Health, the National Institute of Mental Health, and other Federal research agencies. Augmenting Federal efforts are programs supported by private foundations and industry. The cooperation between these agencies and the multidisciplinary efforts of thousands of scientists and health care professionals provide evidence of our nation's determination to conquer brain disease.

To enhance public awareness of the benefits to be derived from brain research, the Congress, by House Joint Resolution 174, has designated the decade beginning January 1, 1990, as the "Decade of the Brain" and has authorized and requested the President to issue a proclamation in observance of this occasion.

Now, Therefore, I, George Bush, President of the United States of America, do hereby proclaim the decade beginning January 1, 1990, as the Decade of the Brain. I call upon all public officials and the people of the United States to observe that decade with appropriate programs, ceremonies, and activities.

In Witness Whereof, I have hereunto set my hand this seventeenth day of July, in the year of our Lord nineteen hundred and ninety, and of the Independence of the United States of America the two hundred and fifteenth.

GEORGE BUSH
Filed with the Office of the Federal Register, 12:11 p.m., July 18, 1990.

Jones and Mendell (1999) have assessed the Decade of the Brain. Noted as a first achievement, was an increase in the visibility of neuroscience. The project provided a forum to publicize brain research. There was an annual Decade of the Brain award given to an important Washington figure who helped raise awareness of the need for brain research. And, there was an annual Brain Awareness Week. The public relations efforts seemed to pay off as bipartisan support for brain research materialized, particularly for NIH. The decade of the project led to a dramatic increase in the number of researchers who identified themselves as neuroscientists with more than one thousand new members being added to the Society for Neuroscience every year of the project.

Beyond the tremendous public relations successes, technical advances that can be attributed to the Decade of the Brain Project include the advancement in brain scanning technologies, advances in brain-computer interfaces (BCIs), progress in developing a vaccine for Alzheimer's disease, the development of tPA to treat ischemic strokes, advances in stem cell research, developments in gene therapy that eventually will lead to a treatment for Parkinson's disease, research showing the existence of neurogenesis, and the development of new drugs.

## The Dana Alliance

Belmonte (2009, p. 2) has described the creation and the goals of the Dana Alliance, an alliance spawned by the private Dana Foundation.

> In 1992, the Dana Foundation decided to share with the public the advances in brain research that were taking place in laboratories and hospitals around the world. After all, the public would be the ultimate beneficiaries of the progress made in knowledge of the brain.
>
> The foundation promoted, first in the United States and later in Europe, the alliance of a group of distinguished neuroscientists who would commit themselves to public awareness of brain research and its potential and to the dissemination of information in a comprehensible and accessible way. Thus was born the Dana Alliance for Brain Initiatives. At that time, many active researchers in the neurosciences regarded the initiative with skepticism, thinking that the responsibility was too big for a private foundation and that it should rather be in the hands of public institutions and governments.
>
> The popularity of Brain Awareness Week, and the success of the multiple publications and activities the world over, including the Dana Alliance's annual progress report, illustrate the error of that judgment. The foundation has succeeded in arousing in the general public the perception that neuroscience directly relates to their personal lives.

In the Spring of 2010 the Dana Alliance hosted its twenty-sixth Learning and the Brain Conference in Washington, D.C. Here as in all previous conferences and publications, neuroscientists presented neuroscience information to non-neuroscientists, achieving the original goals of the Alliance.

## The Human Brain Project

A good description of the Human Brain Project can be found in the RFP (Request for Proposals) issued on April 2, 1993 (available on the NIMH website).

> The general purpose of this initiative (The Human Brain Project) is to encourage and support investigator-initiated basic and clinical neuroscience research and investigator-initiated research on informatics resources that could be used to facilitate neuroscience research. Particular emphasis is placed on research on computer storage and manipulation of neuroscience information, network systems and associated tools that will give neuroscientists access to the stored information. The networks will also provide electronic channels of communication and collaboration to geographically distant laboratories. Emphasis will also be placed on mathematical paradigms linked to empirical research. To optimize the utility of these technologies to neuroscience researchers, they will be developed in the context of specific neuroscience research. It is important to emphasize that the scientific question being addressed is as important as the technology being developed.

The Human Brain Project ran for thirteen years; it stopped accepting grant applications after September 22, 2005. Samson (2006) reported on the debate that emerged with the shutting down of the project. Marcus Heurta, the director of the office of interdisciplinary research and scientific technology at the National Institute of Mental Health (NIMH), said that the program, launched in 1993, had achieved its goal by stimulating the development of advanced technologies for sharing data on the brain through digital databases of brain images, mapping genomic and protein databases, and other data management tools. Heurta added that the Human Brain Project was always meant to be an incubator for neuroinformatics research and that it has been spectacularly successful.

Samson listed the major accomplishments of some forty-two studies supported by the Human Brain Project which were given to him by Heurta.

- 3-D brain atlases, notably of the mouse brain.
- Brain mapping which produced a repository of data on brain structure and function, behavior, and genetic data on seven thousand subjects.
- Cellular recording databases from implanted electrodes.
- Establishment of an fMRI data center which provides computerized analysis of brain images.
- Graphic interfaces that can accommodate varied data types (for example, numerical, textual, graphic, image, and time series).

- New brain-viewing technology that allows neuroscientists in different parts of the world to scan brain images simultaneously and immediately consult with each other.

In response to the criticisms of neuroscientists who believed that the Human Brain Project should not be ended, NIMH replied that there would actually be more research funding available under new NIH initiatives. The difference would be that the future neuroinformatic grant money would be spread across individual NIH Institutes. It is perhaps somewhat unfortunate that the Human Brain Project ran concurrently with the Human Genome Project, also a thirteen year project. The Human Genome Project, completed in 2003, has been viewed as an outstanding success, having identified 99 percent of the 20,000 to 25,000 genes in human DNA. In addition, it has dramatically changed practices in biological research, and in medicine, law, and pharmacology. In defense of the Human Brain Project, however, it is to be noted that unraveling the secrets of the brain is a much more daunting task than is sequencing the human genome.

## The Alzheimer's Disease Neuroimaging Initiative

The five-year Alzheimer's Disease Neuroimaging Initiative (ADNI) project was begun on October 1, 2004, and involved two hundred control subjects, four hundred subjects with mild cognitive impairment, and two hundred subjects with Alzheimer's disease. Information on the Alzheimer's Disease Neuroimaging Initiative (ADNI) can be found on the ADNI website and is described as follows:

> The overall goal of this huge project is to define the rate of progress of mild cognitive impairment and Alzheimer's disease, to develop improved methods for clinical trials in this area, and to provide a large database which will improve design of treatment trials. We expect that this project will provide information and methods which will help lead to effective treatments for Alzheimer's disease, leading to effective prevention.

The ADNI project attracted a great deal of attention not only from researchers but from laity as well. Alzheimer's is not only a disease to be studied, but it is also a concern of many people at large because it is seen as a disease of the aged. This is a concern because the aged are the fastest growing population group in the world. For example, Pollack (2005) noted that today approximately 10 percent of the world's population is over the age of sixty, and that by 2050, this percentage will increase by 100 percent. The greatest increase will be seen among people age eighty-five and over. In the United States in the year 2000, people age sixty-five and over made up 12.3 percent of the population; in 2030, they will make up 19.2 percent of the population. Striking about these statistics is that there is not just growth in the absolute number of adults over sixty-five, but the proportion of the total population that this group occupies will increase as well.

Complementing the above, the June 27, 2009 edition of *The Economist* contained a special report on aging populations. The report noted that rises in life expectancy have been habitually underestimated because it seemed unlikely that the improvement could go on forever but that after each project the numbers had to be revised upward. In the United States in 1940, for instance, there were 3,700 centenarians; today there are over one hundred thousand. The increase is dramatic even after corrections are made for population increases.

More recently emblazoned on the cover of the February 2010 issue of *U.S. News and World Report* are the words, "How to live to 100." In the issue are articles on science and aging, hormone therapy, diet, exercise, working, nature-nurture factors, and Alzheimer's. The implication is that if you live longer the chances of getting Alzheimer's increase. The Alzheimer's article presents seven suggestions that seniors can follow to lower their chances of getting Alzheimer's. One suggestion (p. 49) is to take on mental challenges. Specifically mentioned is the brain's ability to reorganize neural pathways with new information or experiences, meaning that the brain is regularly changing and that we can even generate new brain cells. This statement about the brain carries with it the assumption that Alzheimer's disease can be avoided or delayed by engaging in mental challenges. This issue will be touched on again below.

With increasing longevity can come loneliness and problems with physical health and with finances. However, these problems often take second place to the fear of impending forgetfulness, confusion, bewilderment, Alzheimer's, and dementia. It is no surprise, therefore, that the Alzheimer's project has drawn widespread attention.

## Scanning Technologies

In the late 1970s new imaging technologies were developed. Prominent among them were MRI (Magnetic Resonance Imaging) and fMRI (functional Magnetic Resonance Imaging), the former used to portray structure and the latter, function. In common parlance terms the MRI is often seen as the snapshot or still photo while fMRI is seen as the moving film or video. MRI has been used primarily in the medical arena with images being recorded by radiological technicians and read by radiologists who are physicians. fMRI has been used more frequently in the behavioral research arena with the outputs of this technology being read and interpreted by neuroscientists, most of whom are non-physicians. These technologies advanced the explosive development of neuroimaging because the MRI is superior to the x-ray, and both technologies are non-invasive and non-radioactive. As a result, subjects can be scanned repeatedly within sessions and across sessions. Although the EEG also records brain activity and is also non-invasive, it tells only that there is activity, not its location.

## The Rise of Neuroscience

The new technologies have been the primary drivers of recent neuroscience research. The majority of non-medical studies reported in the literature in the last two decades have used fMRI. Before these modern scanning techniques were developed brain research was usually contingent upon a subject first being a patient diagnosed with some disease or being a patient who suffered some trauma to the brain that required hospitalization and/or brain surgery.

The new technologies opened the door for non-physicians to do neuroscience research. As noted above, during the Decade of the Brain Project there was an increase of one thousand neuroscientists per year for each of the project's ten years. Members of the new cohort came from a number of disciplines including biology, computer science, psychology, and sociology. With the widened diversity in the population of neuroscientists, new professional associations were formed and new neuroscience journals followed in tow.

For example, MIT Press and the Cognitive Neuroscience Institute published the first volume of the *Journal of Cognitive Neuroscience* in 1989. The purpose of the journal was to investigate brain-behavior interaction and to promote lively interchanges among mind scientists. In 1994 the Cognitive Neuroscience Society was formed. A relationship between the cognitive neuroscience journal and the society was formed. In 2008 the first issue of *Social Neuroscience* was published by Psychology Press. The goal of this journal was to provide a place to publish empirical articles that intend to further our understanding of the role of the central nervous system in the development and maintenance of social behaviors. Formation of the new journals and the new society announced to the world that the neuroscience tent had been made larger and that its inhabitants were no longer under the fabric only of medical personnel.

Other journals in the field would include *Aging, Neuropsychology and Cognition, Aphasiology, Applied Neuropsychology, Brain Injury, Child Neuropsychology, The Clinical Neuropsychologist, Cognitive Neuropsychiatry, Developmental Neuropsychology, Journal of the History of Neuroscience, Journal of Clinical and Experimental Neuropsychology, Neurocase,* and *Neuropsychological Rehabilitation.*

The expansion of the neuroscience field can also be seen in universities. In the beginning, when neuroscience expanded beyond the medical arena, non-medical neuroscientists were housed in a number of departments, particularly in biology and psychology departments. Now every major university has a neuroscience department or program. For example, the Association of Neuroscience Departments and Programs (consolidating with the Society for Neuroscience in July 2009) listed 203 departments and programs dedicated to the study of neuroscience. The departments were named neuroscience departments. The nomenclature, seemingly trivial at first glance, permits the departments to be multidisciplinary. No single discipline is privileged.

## Interest by the Public at Large

In October of 2007 the magazine, *Wired,* published a supplement titled the *Wired Geekpedia.* The supplement was designed as a high tech encyclopedia that provided descriptions of new terms and concepts in the modern digital age. One section (p. 35) in the supplement dealt with the prefix, *neuro. Wired* noted that this prefix was becoming one of the most popular prefixes and listed thirty-seven suffixes to which this prefix has been attached. Included in the list of concatenations were neuroscience, neuroanatomy, neurophysiology, neurofeedback, neurofitness, neurobics, neurobiology, neurochemistry, neuroendocrinology, neuropharmacology, neuroceuticals, neuropsychology, neuropsychiatry, neurotechnology, neuroinformatics, neuroimaging, neurocomputation, neuroengineering, neuromacy, neuromorphics, neurocybernetics, neuroweapon, neurosociety, neuropolicy, neuroeconomics, neurofinance, neuroforecasting, neuromarketing, neurophilosophy, neurophenomenology, neuroesthetics, neuroergonomics, neurotheology, neurolinguistics, neurosemantics, neuroethics, and neurosecurity. This robust list of concatenations gives further evidence that this is indeed the age of the brain.

Some manifestations of the expanded awareness of the brain in everyday life can be seen in recently published books. For example, there is *This Is Your Brain on Music* (Levitin 2006) which holds that there is a genetic basis for the universal love of music; *Your Money & Your Brain* (Zweig 2007) which tells its readers how neuroeconomics can make them rich; *The Neuroscience of Fair Play* (Pfaff 2007) which tells its readers why people usually follow the golden rule; *buy.ology* (Lindstrom 2008) which claims to report on the largest neuromarketing project ever conducted; and *How We Decide* (Lehrer 2009) which is a treatise on the neurological underpinnings of decision making. The fact that these books were written, and the fact that there was an audience for the books, give testimony supporting the assertion that this is the age of the brain. An outlier in brain books is *Kluge* (Marcus 2008). Marcus takes the position that the brain is a klutzy, inelegant organ. It forgets, its perceptions can change even though the stimulus is fixed, and it communicates ambiguously, vaguely, and deceptively. Computers would never behave this way. But Marcus admits that these many "inadequacies" of the brain also allow humans to be very creative. Think just of the creative uses of the metaphor, for instance.

The accelerated study of the brain produced results that were encouraging to the public at large, and many recent books testify to this encouragement (Kiefer 2007, Larson 2008, MacDonald 2008, Medina 2008, Anthes 2009, and Restak 2009). In general, these researchers reported that what is good for the body is good for the brain, that the "use it or lose it" rule applies to the brain as well as the body, and that the brain needs to be stimulated if cognitive skills are to be maintained.

One harbinger of things to come can be seen in the report of Helman (2008). Helman observed T. Boone Pickens, an eighty-year-old Texas multibillionaire, submit to a brain scan at the University of Texas at Dallas. Pickens, who donated eleven million dollars to the center, agreed to be a subject for the university's study on brain function and age. Pickens came through with flying colors, displaying brain functioning similar to a fifty-five-year-old person rather than an eighty-year-old on easy decision questions, hard decision questions, and in judging distances. While on the surface Pickens' volunteering seemed like an act of generosity, the subtext showed that here was an eighty-year-old who still had "all his marbles." His volunteering perhaps set the bar for future elderly people in responsible positions to have their brains scanned, and to have their scans made public. It is worth mentioning that Pickens' general life style supports the assertions made at the end of the previous paragraph. Helman said, "Pickens does all the right things to keep a brain healthy: trying new activities and exercising regularly. On mornings when Pickens is in Dallas he works out with his trainer. He's never smoked, doesn't drink coffee, and hasn't touched a glass of Scotch in six years."

One approach to stimulating the brain cognitively is the use of computer games. Mossman (2009) has reviewed a number of computer games for their ease of use and the degree to which they motivate the user. An important point raised by Mossman is that many of the programs present the standard fare games and in this sense do not differ from one another. Instead, the style of presentation is instrumental in ranking them. For example, some games have a coach or trainer who greets the user with each log on and states what the user has accomplished and what still needs to be achieved. Other games sense that a plateau has been reached and let the user relax before proceeding with more difficult tasks.

Can playing computer brain games help the brain? The evidence says that they can, although the evidence is still tentative with respect to the maintenance and enhancement of cognitive skills. Green and Bavelier (2006) say, "Playing action-video games can alter fundamental characteristics of the visual system." And, "Essentially, video games provide a convenient and easily accessible means for changing the brain."

According to Granek et al. (2007), "The inherent brain mechanisms for performing complex skills are different in highly experienced video gamers." And, "Video-game training reorganizes the brain's activity and leads to more efficient and effective control of skilled movements other than playing video games."

Finally, Restak (2009), a neuroscientist and neuropsychiatrist, said:

> . . . if you take up video-gaming, even casually, you can expect the following benefits: decreased overall reaction times, increased eye-hand coordination, and enhanced manual dexterity. You'll improve your spatial visualization skills and your ability to mentally work in three dimensions. Finally, you'll be better able to divide and rapidly switch your attention as well to increase the number of things that you visually attend to simultaneously. (p. 166)

And:

> While no one has convincingly demonstrated that video game playing can delay or prevent the onset of dementia, there is general agreement that short-term memory and speed of response times are enhanced among video gamers. The decrease in reaction time is especially intriguing and raises an interesting and so far unanswered question: Does IQ increase in tandem with reaction time improvements? (p. 160)

## Summary

Interest in the brain has migrated from the surgical theaters of the past to contemporary scanning laboratories, and now to the public at large. The Decade of the Brain and The Human Brain Project produced many research findings with great heuristic value, that is, they raised more research questions which in turn stimulated more research. But perhaps more importantly they, along with the Dana Alliance, created a greater awareness of the importance of the brain in the minds of key policy makers, particularly in the United States Congress. The Alzheimer's Disease Neuroimaging Initiative was a response to a national fear particularly among the elderly. One of the greatest fears among the elderly is the fear of Alzheimer's disease and dementia. Much progress has been made in mending the body with hip and knee replacements and non-implanted prosthetic devices, but little progress has been made in preventing Alzheimer's disease. The fear of Alzheimer's by the public at large has prompted a number of books and computer games designed to help people maintain and enhance their cognitive skills.

The genie is out of the bottle. The momentum has been established. From the surgical theaters, the incubated neuroscience laboratories, the neuroscience university programs, the publishers of serious books, to the public at large, this truly is the age of the brain.

# *Chapter 2*
# Neuroscience Notes

The purpose of this chapter is to present several quotes testifying to the complexity of the brain, to provide a brief history of neuroscience, to present some basic information on the functioning of the brain and its physiology, and finally to present some basic geographical markers of the brain's anatomy. The thumbnail historical sketch will be presented in a chronological collage of some of the most prominent contributors to our knowledge of the brain. The collage is also an obituary of sorts since all of the actors listed have died. The names of living and current contributors to the field will emerge in subsequent discussions.

Testifying to the complexity of the brain are a poet, Emily Dickinson, a philanthropist (Allen 2009), and a neuroscientist (Gazzaniga 2008).

## Complexity of the Brain

### The Poet

*The brain is wider than the sky,*
*For put them side by side,*
*The one the other will include*
*With ease, and you beside.*

*The brain is deeper than the sea,*
*For, hold them, blue to blue,*
*The one the other will absorb,*
*As sponges, buckets do.*

*The brain is just the weight of God,*
*For, lift them pound for pound,*
*And they will differ, if they do,*
*As syllable from sound.*
*(Emily Dickinson)*

*The Philanthropist*

The mystery of how the brain works is the most compelling question in science. We can discover new planets around distant stars and find water on Mars, but over 95% of the workings of the brain remain unexplored and unexplained.

*(Paul Allen)*

*The Neuroscientist*

The human brain has approximately 100 billion neurons, and each, on average, connects to about 1,000 other neurons. A quick multiplication reveals that there are 100 trillion synaptical connections. So how is all this input getting spliced and integrated into a coherent package? How do we get order out of this chaos of connections? Even though it may not always seem so, our consciousness is rather kicked back and relaxed when you think about all the input with which the brain is being bombarded and all the processing that is going on. In fact, it is as if our consciousness is out on the golf course like the CEO of a big company while all the underlings are working. It occasionally listens to some chatter, makes a decision and then is out sunning itself.

*(Michael Gazzaniga)*

# History

One way to describe history in any area is to list the key players in the area over the years. Such a history in neuroscience would undoubtedly include the following names. The list is also a necrology for each person named has died. Noted living neuroscientists are not ignored and will be referenced in other parts of this book. The chronology below reflects the transitions reported in chapter 1. Seen is that early neuroscience writings and discoveries were made by philosophers and physicians, and, with time, the participant population became more inclusive, welcoming biologists and psychologists, for example. Many of the "offspring" of the new participants now call themselves neuroscientists as they shed their old discipline memberships.

*Hippocrates (circa 460 BC–circa 370 BC)*: Penfield and Roberts (1966, p. 7) quote the words of Hippocrates.

> Some people say that the heart is the organ with which we think, and that it feels pain and anxiety. But it is not so. . . . Men ought to know from the brain

> rows, pains, griefs, and tears. Through it, in particular, we think, see, hear, and distinguish the ugly from the beautiful, the bad from the good, the pleasant from the unpleasant. . . . To consciousness the brain is the messenger. For when a man draws breath into himself, the air first reaches the brain, and so is dispersed through the rest of the body, though it leaves in the brain its quintessence, and all that it has of intelligence and sense. . . .
>
> Wherefore I assert that the brain is the interpreter of consciousness. . . .
>
> To consciousness the brain is the messenger. . . . The brain is the interpreter of consciousness.

Hippocrates, with the exception of his beliefs on respiration, was right on target with his assertions on the brain. This is notable since his was a lone voice at the time he made his assertions.

*Aristotle (384-322 BC)* believed that the heart was the seat of perception, cognition, and expression. The brain's function was to cool the passions of the heart.

*Claudius Galen (circa 130–200 AD)* the Greek physician, Galen, believed that the heart was the seat of the soul or mind. However, he believed that vital spirits arose from the heart and were made noble in the brain. During his time Galen was considered to be the authority on medicine.

*René Descartes (1596–1650)* was the famous French dualist who believed that the mind or soul was immaterial and was distinct from the physical body. He believed that the mind and the body intersected in the pineal gland.

*Franz Joseph Gall (1758–1828)*, the German anatomist, believed that certain mental functions are localized in the brain, and as these functions develop, the changes are manifested in topographical "bumps" on the outer skull. Although Gall's views were later discredited, he did focus attention on the possibility of localized functions in the brain. Much of today's neuroscience research that seeks answers to questions of brain and behavior is focused on localizing functions. The major difference is that today's researchers have the technology to localize function inside the skull which Gall, of course, did not.

*Paul Broca (1824–1880),* a French physician, observed a patient who could understand spoken language but who could not speak fluently. Broca found on autopsy that there was a lesion in the anterior portion of the temporal lobe of the left hemisphere of the brain. The observed disorder is now called Broca's aphasia or expressive aphasia.

*Camillo Golgi (1843–1926)*, Nobel Laureate, was the developer of the silver staining method (now known as the Golgi stain). The staining method is used primarily to highlight the dendrites and cell bodies of a neuron.

*Karl Weigert (1845–1904)* developed a stain for the myelin sheath surrounding the axons of neurons. This is often used with a contrasting stain to show the distinctions between the neurons' cell bodies and their axons.

*Karl Wernicke (1848–1905)* described a language disorder where a patient could speak but could not understand language. Patients suffering from this disorder, known as Wernicke's aphasia, showed brain damage in the posterior por-

order, known as Wernicke's aphasia, showed brain damage in the posterior portion of the temporal lobe of the left hemisphere. That area of the brain is known as Wernicke's area.

*Santiago Ramon y Cajal (1852–1943)* used Golgi stains to show that the neuron was the basic unit of the nervous system. He shared the Nobel Prize with Golgi in 1906.

*Charles Sherrington (1857–1952)* was responsible for naming the junction between neurons, the "synapse." He shared the Nobel Prize with Edgar Adrian in 1932.

*Franz Nissl (1860–1919)* discovered that certain dyes were effective in staining neurons, particularly their cell bodies and proximal dendrites.

*Korbinian Brodmann (1868–1918)* used fellow German Franz Nissl's staining techniques to map the human brain. The cortex was mapped into about 50 different areas. Brodmann's system is still widely used today. Brodmann's areas are usually designated with BA followed by the area number.

*Karl Lashley (1890–1958)* is best known for the principle of *mass action*, which states that the cerebral cortex acts as one—as a whole—in many types of learning and for the principle of *equipotentiality*, which states that if certain parts of the brain are damaged, other parts of the brain may take on the role of the damaged portion.

*Antonio Moniz (1874–1955)*, a Nobel Laureate, was the developer of the frontal lobotomy surgical technique, a technique which has fallen out of favor.

*Wilder Penfield (1891–1976)*, in collaboration with Lamar Roberts and Herbert Jasper, operated on patients with severe epileptic seizures. After exposing the brain and before beginning surgical procedures, Penfield would electrically stimulate various regions of the brain while the patient was awake and conscious. This permitted him to gain information on the minimum amount of brain tissue that needed to be surgically removed to provide symptom relief, and it provided the information necessary to create maps of the brain showing the brain structures involved in various functions, particularly speech and language functions. Penfield studied with Sherrington in England before practicing in Canada at the Montreal Neurological Institute.

*Donald Hebb (1904–1985)*: A neuroscientist at McGill University, put forward the notion that learning involves the strengthening of the connections of preexisting neurons rather than the formation of new synapses. He often used the metaphor of the small village where neurons were the villagers, fighting fires at one time, debating school improvement at another time, celebrating a holiday at yet another time, and so on.

*Roger Sperry (1913–1994)*, a Nobel Laureate, is famous for testing ten patients who had undergone surgery that severed the corpus callosum, the "300 million lane expressway" that connects the hemispheres. Sperry and his associates designed tasks that were hemisphere specific and contributed significantly to the brain lateralization literature.

*James Olds (1922–1975)*, working at a number of major universities in the United States, is credited with identifying the pleasure centers in the brain using electrical stimulation to the brain. Working primarily with rats in Skinner boxes, Olds found that electrical stimulation to various regions in the hypothalamus of the rat resulted in the rat's repeating the behavior that immediately preceded the shock. These results suggested that the mild electrical shock was a reward in the operant conditioning paradigm, supporting the behaviorist's law of effect.

# Micro Building Blocks of the Brain

## The Neuron

The principal brain cell is the *neuron.* Neurons range in size from a millimeter to several feet in length (those that go from the spinal cord to the toes, for example). Figure 2.1, in the photo spread at the end of the chapter, shows a schematic diagram of a neuron. At one end is the cell body with *dendrites* projecting from the cell body. The dendrites are receivers; they receive messages from other neurons. The dendrites are attached to the neuron's cell body. Extending prominently from the neuron cell body is the axon. The axons are senders; they send messages to other neurons. At the beginning of the axon is the *axon hillock* a bulbous "bump" that serves as a sort of gatekeeper, determining if the inputs to the neuron are sufficient to cause the neuron to fire. If the neuron does fire, an action potential travels down or along the axon to the other or terminal end of the axon, the *axon terminals.* The movement of the action potential through the axon is facilitated by the *myelin sheath,* the "insulation" material surrounding the axon. The myelin sheath increases the speed of movement of the action potential, and it prevents electrical discharges in one neuron from disturbing neuronal activity in non-targeted areas.

## Neuron to Neuron Communication

Unlike the computer, to which the brain is often compared, neuron-to-neuron transmission in the brain is electrochemical rather than purely electrical. Communication between neurons is accomplished by the transmission of a chemical, called a *neurotransmitter*, across a *synapse*, a space that separates two neurons. Communication across the synapse, then, is chemical. Figure 2.2, in the photo spread at the end of the chapter, shows this relationship schematically. The presynaptic neuron releases a neurotransmitter through the axon terminals. The neurotransmitter is then absorbed by the receptors on the dendrites of the postsynaptic neuron. There are many neurotransmitters, some of which have perhaps not yet been identified. Two major categories of neurotransmitters are the excitatory and inhibitory. The former increase the chance of the postsynaptic firings, and the latter decrease the chances of firing.

After the neurotransmitter is released into the synapse and absorbed by the postsynaptic neuron, some neurotransmitter may still remain in the synaptic gap. This "excess" neurotransmitter can be broken down by surrounding enzymes or can be reabsorbed back into the presynaptic neuron.

Some common neurotransmitters are:

- Acetylcholine: Triggers muscle contractions and is involved in memory, anger, and aggression. Lack of acetylcholine plays a part in memory lapses associated with Alzheimer's. Cholinesterase inhibitors used to treat Alzheimer's prevent the breakdown of acetylcholine.
- Dopamine: Helps control movement, modulates mood, motivation, and reward. Drug addicts use drugs to boost its effect. Dopamine levels are low in Parkinson's sufferers. The drug, Levodopa (L-dopa), elevates dopamine levels and, as a result, can improve the movement problems of Parkinson patients.
- GABA: Involved in movement and regulation of anxiety; moderates neuron firing. Low levels of GABA lead to floods of brain signals, causing epileptic seizures. Treatments for epilepsy and anxiety work by increasing GABA transmission.
- Glutamate: Closely associated with memory and learning; it's the main excitatory transmitter and is thought to have a role in Alzheimer's.
- Norepinephrine: Plays a part in stress responses; influences alertness, arousal, and reward. Implicated in anxiety and mood conditions, including depression and bipolar disorders. SSRIs and other antidepressants enhance transmission of norepinephrine.
- Serotonin: Helps regulate mood, body temperature, sleep, and appetite. Depression, impulsive behavior and aggression are linked to serotonin. SSRIs (Selective Serotonin Reuptake Inhibitors), which improve serotonin transmission are often taken to treat depression. Prozac operates by blocking the re-uptake of serotonin. It is an SSRI drug that was developed in 1987. SSRIs are a class of antidepressants used in the treatment of depression, anxiety, and some personality disorders. SSRIs inhibit the reuptake of the neurotransmitter, serotonin, back into the presynaptic cell. As a result there is more serotonin available to bind with the postsynaptic receptor neuron.

In general, then, the effects of many pharmacological drugs, legal and illegal, are a result of their impact on the brain's neurotransmitters. Sapolsky (2005) provides some additional examples:

- Hallucinogens such as LSD, mescaline, and psilocybin are able to artificially stimulate the serotonin receptors.
- Curare can block the acetylcholine receptors in the diaphragm, causing breathing to cease.

- Antipsychotic drugs block the dopamine receptors, lessening symptoms of schizophrenia, for example.
- Amphetamines can trigger the premature release of dopamine transmitters.
- Because the release of dopamine makes a person feel pleasure in at least one part of the brain, drugs such as cocaine become highly addictive.
- Drugs that release dopamine can trigger schizophrenia while dopamine blockers can inhibit schizophrenic behavior.
- Destroying the neurotransmitter norepinephrine can lower blood pressure.

## Glial Cells

Neurons have royalty status while glial cells have been the commoners in the kingdom of the brain. Seen for many years as serving only a supportive function, now the number of roles recognized as being played by glial cells is increasing. Fields (2011a) has reported his own work and has reviewed the work of others who have studied glia. Fields first notes that neurons occupy only 15 percent of our brain cells; glial cells make up the rest. For example, glial cells called astrocytes carry nutrients and waste and are involved in neuronal communication, glial cells called oliogodendrocytes insulate axons, increasing transmission speeds, and microglial cells promote repair when the brain is injured. When the microglia fail so does the brain.

Fields notes that for one hundred years neuroscience has operated on an idea called the neuron doctrine. This doctrine holds that all information in the nervous system is transmitted by electrical impulses over networks of neurons connected by synaptic connections. But this idea is flawed, according to Fields, because some information transmitted in the brain bypasses neurons completely. It is transmitted by the glia. Moreover, neurons and glia work together in a variety of ways. The new and forthcoming information on glia suggests that a paradigm shift has occurred in neuroscience. More discussion of the glial cells will be reported in the methodological chapter in part 3 of this book.

# Macro Structures in the Brain

Figure 2.3, in the photo spread at the end of the chapter, shows the *cerebral cortex,* which is the thin outer layer of cells in the brain. The cerebral cortex is the most complex area of the brain. It is the seat of memories, intellect, reasoning, creativity, and intelligence. The cerebral cortex covers both sides of the brain. It is wrinkled and folded, masking its overall area which would be about two and one half square feet if it were flattened. The folds in the cortex are called *gyri,* and the valleys or fissures are called *sulci*. The thickness of the cerebral cortex varies between one and four millimeters. The deep sulci are used to mark off regions or lobes of the cortex. Figure 2.3 shows, in clockwise fashion, the *pre-*

*frontal*, *frontal*, *parietal*, *occipital*, and the *temporal lobes*. Each lobe is represented on both sides of the head giving rise to the description of cerebral hemispheres, left and right. The cerebellum, portrayed on the figure, will be discussed below.

The *left hemisphere* has often been called the *dominant hemisphere* because in about 97 percent of the people in the world the left hemisphere is dominant for logical thought and for the reception, analysis, and production of speech and language. On the reception side there is an area in the left hemisphere at the temporal-parietal lobe border called Wernicke's area (named after the physician, Karl Wernicke: see above) responsible for the decoding of speech and language. Damage to this area impairs the sufferer's ability to understand speech. At the lower anterior portion of the temporal lobe is an area called Broca's area, named after the physician, Paul Broca (see above). This area is responsible for encoding speech. In general, damage to Wernicke's and Broca's areas or to neighboring areas often results in a range of combinations and severities of disorders of comprehension and production of spoken and written language.

The *right hemisphere* has been called the minor hemisphere from the fact that language has been given a privileged status. It is often referred to as the nonverbal hemisphere, responsible, for example, for perception, intuition, and the recognition of faces, melodies, and time. Roger Sperry (1975), who won the Nobel Prize for his split brain experiments, has said that each hemisphere has its own way of learning and expressing: the left through speech and language and the right through nonlinguistic perceptions and messages. Sperry claimed that the right hemisphere was being discriminated against because little teaching was aimed at the right hemisphere. There have been some responses to Sperry's claims: principally, the publication of books on emotional and social intelligence and innovation. At the same time the right brain-left brain dichotomy has become somewhat of a metaphor. Artistic people are often labeled as right brained and the number crunching scientific types as left brained. While there may be some validity behind these labels the process is often an over-simplification. In addition there is some overlap in the functions of the two hemispheres. And finally, when one area of the brain is diseased or surgically removed, the intact side often begins to take over the duties of the diseased or excised tissue on the other side. This is especially true if the insult occurs at an early age.

The *frontal lobes* (see Figure 2.4 in the photo spread at the end of the chapter) of the brain are the most recently developed portions of the brain in terms of evolutionary development. They are involved in executive functions. In general they are involved in planning and executing complex behaviors, in the expression of one's personality, exhibiting proper social behavior, and in decision making. They analyze the present and make judgments about what actions to take as a consequence of the analysis. They make moral judgments: what is good and what is bad. When there is damage to the frontal lobes there is often a change in personality. A person who was socially adept before disease or damage to the frontal lobes may become a social disaster after the disease or damage. Imagine the public behavior of the worst drunk you know, and this will give you some idea of what the behavior of a person with frontal lobe pathology may exhibit.

The anterior parts of the frontal lobes are called the *prefrontal lobes*. The *dorsolateral prefrontal cortex* (DL-PFC) is the last area of the brain to develop in humans. Not until the early twenties is this area fully developed. This fact has led to the Allstate Insurance Company's running a print ad which asks: "Why do most sixteen-year-olds drive like they're missing a part of their brain?" "Because they are." (This ad has appeared in a number of magazines. See the January/February 2010, issue of *The Atlantic* for an example). The DL-PFC is instrumental in intellectual functioning and action, and integrating sensory information with memory information. The *medial prefrontal cortex* (MPFC) seems to be highly involved in emotion regulation. It integrates cognition and affect, for example, and appears to integrate emotional memories with the present.

At the back of the frontal lobe is the *primary motor area* (represented in Figure 2.4 in the photo spread at the end of the chapter). The primary motor area runs up, over, and down the left hemisphere and then up, over and down the right hemisphere of the brain. The primary motor area directs the movements of the body. Since the large motor tracts that descend from the brain cross over to the opposite side of the body, the left side of the primary motor area controls the right side of the body and vice versa. Common evidence of this crossover can be seen in many stroke patients who have severe problems with speech and language (implying that the trauma was suffered by the left side of the brain) and paralysis and/or weakness on the right side of the body.

The *parietal lobes* (one on each side of the head) are located behind the frontal lobes toward the top of the head and above the ears. The parietal lobes are concerned primarily with sensation. Right behind the primary motor strip at the back of the frontal lobe is the *sensory area* at the front of the parietal lobe. This area of the parietal lobe receives sensory messages about touch, pain, and temperature. Similar to the contra-lateral connections emerging from the primary motor area, the primary sensory area responds to sensations on the opposite side of the body. Directly behind the sensory strip in the parietal cortex is the *sensory association area* (not noted in Figure 2.4, in the photo spread at the end of the chapter).

Here is where the complex facets of sensation are integrated. For example, in a testing situation if one were asked if the red block is closer to the green block than it is to the orange block, the right side of the parietal lobe would have to "kick in" to answer the question. Much integration would be involved.

Continuing in clockwise fashion, at the back of the head are the *occipital lobes*, one on each side of the head. The occipital lobes are concerned with vision. We think of the eyes when we think of vision, but the eyes only collect visual information. It is at the very back of the occipital lobes in an area called the primary visual cortex where what we see is interpreted. In the occipital lobes, as in the frontal and parietal lobes, the wiring to the periphery (in this case the eyes) is crossed. The left visual cortex receives stimuli from the right half of

inverting lenses. Inverting lenses turn the world upside down and indeed when subjects first put on the inversion lenses the world is literally upside down. But gradually over time the world becomes righted, showing the capacity of the primary visual cortex to change or interpret what we see.

Finally the *temporal lobes*, one on each side of the head, are located beneath the frontal and parietal lobes and in front of the occipital lobes. The temporal lobes are important in memory, speech, language, hearing, imagination, dreaming, and daydreaming, for example. While it was once believed that virtually all long term memories were stored in the temporal lobes, this view has been tempered, for now it seems that many areas of the brain are involved in storing long term memories. In fact, memories may have multiple storage sites, something referred to as redundant memory storage. Penfield and Roberts' (1966) experiments, in which various parts of the brain were electrically stimulated, demonstrated that the temporal lobe is certainly involved in memory. They found that only in the temporal lobes (right and left) did electrical stimulation trigger old memories. Often the triggered memories seemed to be quite benign, seeing someone standing at one's desk with a pencil, for example. What surprised Penfield and Roberts were the details that were remembered even in what seemed to be innocuous situations. This seems to stand in contrast to our knowledge of the effects of emotion on memory. Highly charged emotional events are branded on our memories forever. Of course, forgetting is often less of a memory problem than it is an access problem. If someone were asked what the capital of Canada is, the answer may not be forthcoming. However, if one were given the choices, Montreal, Toronto, Ottawa, or Vancouver, the person questioned might immediately say, "Of course, it's Ottawa." The answer was there; it was simply a problem of access.

Other neuron concentrations listed on Figure 2.4, in the photo spread at the end of the chapter, are, again in clockwise fashion, the *nucleus accumbens*, the *medulla*, the *pons*, the *hippocampus*, the *amygdala*, the *cerebellum*, the *thalamus*, the *basal ganglia*, the *corpus callosum*, the *cingulate gyrus,* the *supplementary motor area*, and the *anterior cingulate gyrus*. Each half of the brain has one nucleus accumbus. The nucleus accumbus is thought to play a key role in reward, addiction, laughter, pleasure, fear, and the placebo effect. The medulla is involved with autonomic functions such as heart rate, breathing, and blood pressure. The pons is a structure located on the brain stem. It functions to send signals to the cerebellum and medulla and it transmits sensory information to the thalamus. It is also involved in sleep, respiration, swallowing, bladder control, hearing, equilibrium, taste, facial sensation, posture, eye movement, and facial sensations.

The hippocampus is a paired structure located beneath the temporal lobes of the brain. It plays important roles in memory and spatial orientation. The amygdala is involved with emotion, particularly fear. In some cases the amygdala will trigger an immediate response. For example, if a child darts out in front of your car, you would take immediate action by hitting the brakes or swerving away from the child. You would take no time to rationally analyze the situation. For instance, you would not think, is that really a child or is that something that looks like a child? Instead, you would take immediate evasive action. It is not surprising that the amygdala is involved in memory since emotion and memory are yoked. In many cases it is one-trial learning. The child who touches the hot stove seldom if ever touches it again.

The cerebellum, which looks like a little brain behind the brain, plays an important role in movement. It does not initiate movement, but it coordinates movement, leading to precise, coordinated, and accurate movements. Think of the football quarterback who accurately throws the ball not to the receiver, but to where the receiver will be. This requires the involvement of the cerebellum's interaction with the visual system. The thalamus has been called the great sensory weigh station. It "evaluates" and relays sensations to the motor cortex. Other but lesser functions are its involvement in relaying motor signals to the cortex along with its involvement in regulating consciousness, sleep and alertness. Influenced by many parts of the brain, the basal ganglia are involved in decision making. Specifically, the basal ganglia are instrumental in deciding which of several possible actions to take at a given time. The basal ganglia can be compared to a policy-making body which discusses which of a number of policies to instate in a particular situation.

The corpus callosum has been called that 300 million lane expressway connecting the two hemispheres. It indeed is the main communication pathway between the hemispheres. It was the corpus callosum that was severed (in an effort to relieve the effects of severe epilepsy) in the patients seen by Sperry in his split-brain experiments. The cingulate gyrus is involved with emotion, learning, and memory. The anterior part of the cingulate gyrus seems to be involved in suppressing inappropriate actions. Finally, the supplementary motor area (SMA), which is located anterior to the primary motor cortex, differs from the primary motor area in that the SMA is involved in triggering actions that are prompted from memory and not from external cues.

## Caveats

The above brief review of some of the areas and structures of the brain, identified in Figures 2.3 and 2.4 is, at most, only a thumbnail sketch. First, only a few of the areas and structures in the brain have been identified. Second, we must keep in mind the complexity of the brain and how little we know about its structure. More and more research has shown that there seem to be many more interconnections among structures than we had imagined. If and when these suspected increasing connections are confirmed, then we may see that the brain works much more as a whole than previously thought. For example, memory was once seen to be located almost solely in the temporal lobe, but today memory is seen as being represented all over the brain. Again, what we have briefly outlined above are the primary functions of various areas and structures of the brain.

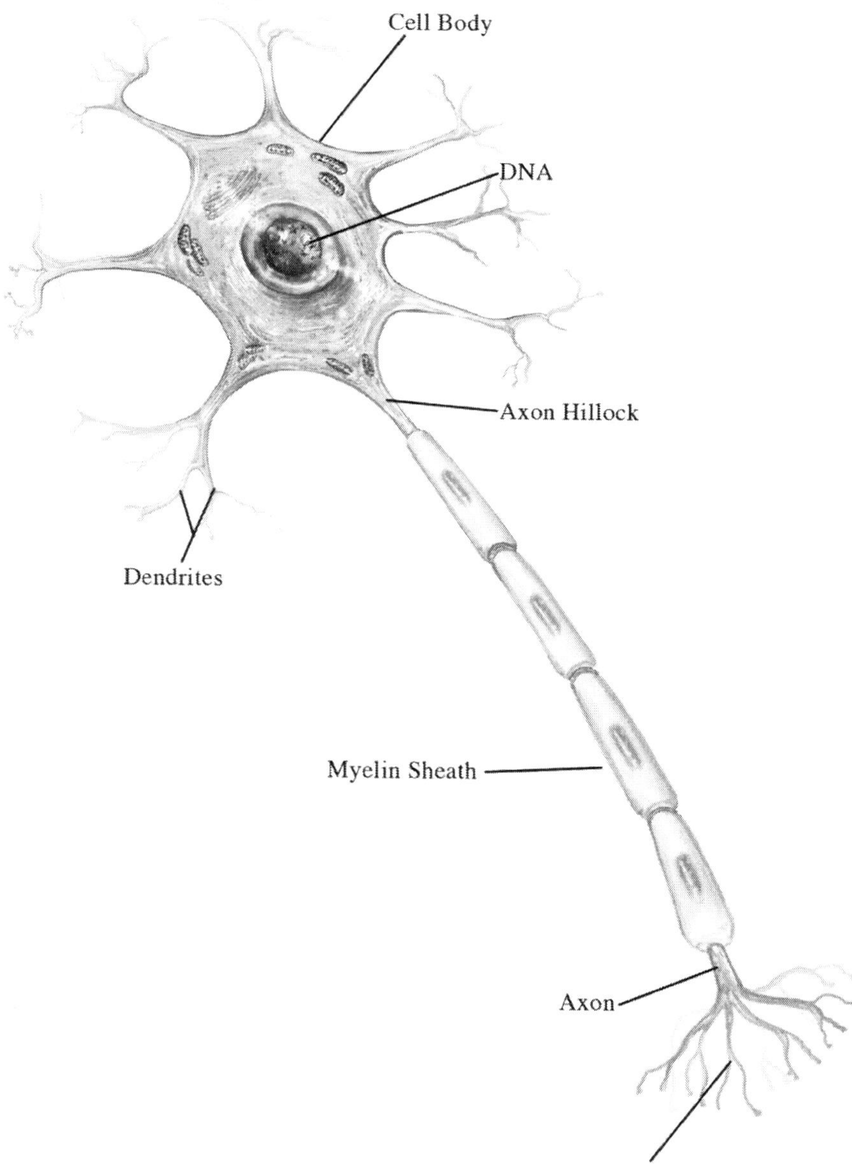


***Figure 2.1.*** *A graphic representation of the neuron from its dendrite message receptors to its transmitting axon terminals. Image courtesy of Randal McKenzie.*

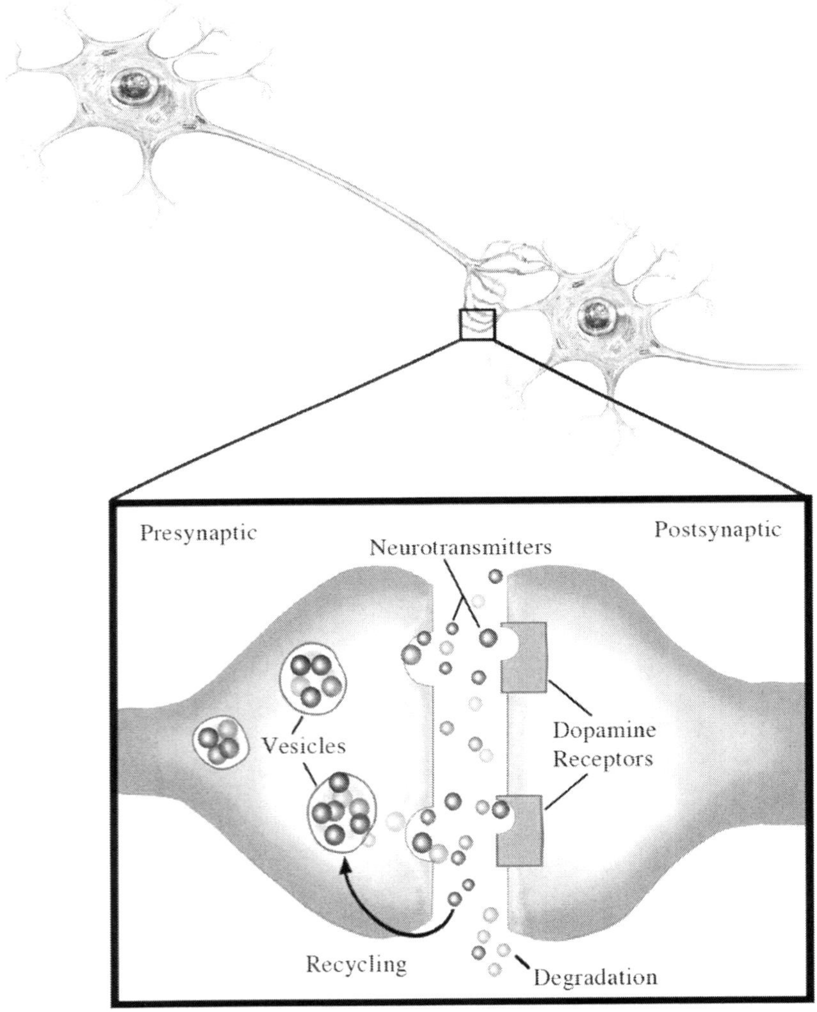


***Figure 2.2.*** *Communication between neurons as accomplished by the transmission of a chemical, called a neurotransmitter, across a synapse, a space that separates two neurons shown schematically in the figure. Image courtesy of Randal McKenzie.*

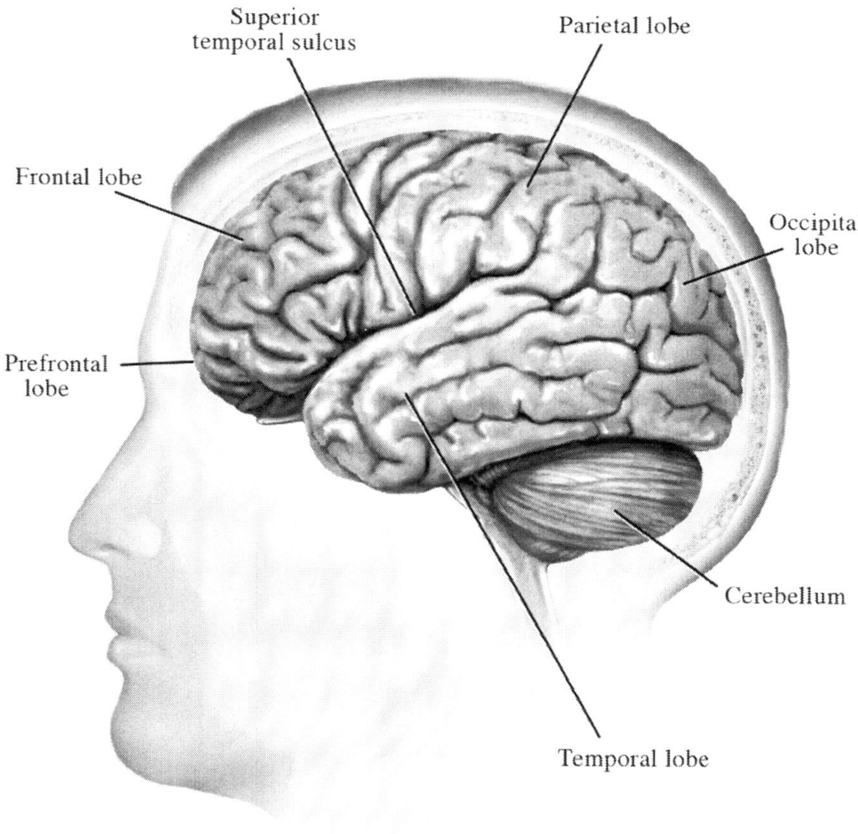


**Figure 2.3.** *Divisions of the cerebral cortex, commonly called lobes, which are the outermost layers of the cerebral hemispheres. Image courtesy of Randal McKenzie.*

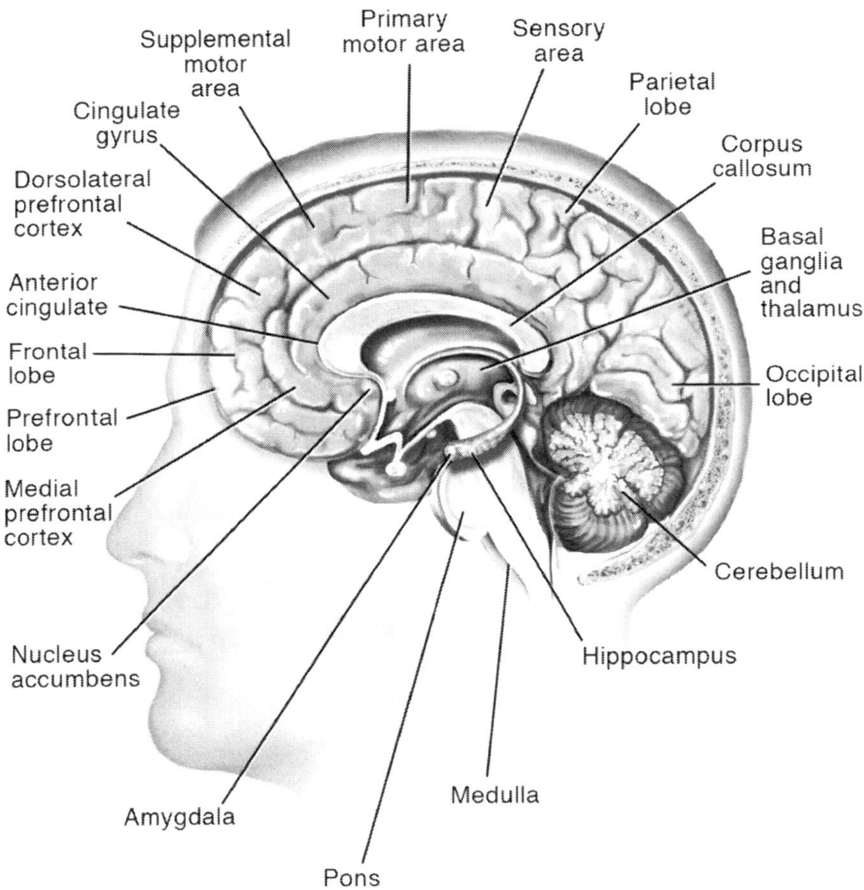


***Figure 2.4.*** *A side view of the brain showing selected cortical and sub-cortical areas. Image courtesy of Randal McKenzie.*

*Chapter 3*
# Methodologies

## Notes on Scientific Progress

It has been noted that science progresses one funeral at a time. This means that any scientific finding is tentative and stands to be confirmed, revised, or overturned completely. Bertrand Russell has said that the whole problem with the world is that fools and fanatics are always so certain of themselves, but wiser people are so full of doubts (see www.quotationspage.com). More dramatic is the statement of Jacob Bronowski (1973, p. 374):

> Science is a very human form of knowledge. We are always at the brink of the known, we always feel forward for what is hoped. Every judgment in science stands on the edge of error, and is personal. Science is a tribute to what we can know although we are fallible. In the end the words were said by Oliver Cromwell: 'I beseech you, in the bowels of Christ, think it possible you may be mistaken.'

This preamble is given because any findings reported, including those given in this book, are tentative and can be confirmed, revised, or overturned in the future. For example, for years the dictum was that adults can only lose neurons, that no new neurons are formed in the brain. Recent research, however, suggests that the brain produces new neurons, particularly in the hippocampus, a key structure involved in memory. A long held dictum has apparently been overturned.

The tentative nature of scientific results was made manifest in an article by Ioannidis (2005) titled, *Why Most Published Research Findings are False*. Ioannidis analyzed forty-nine of the most highly regarded findings in medicine over a thirteen-year period. Of the forty-nine articles, thirty-four were re-tested and fourteen of these, or 41 percent, were shown to be wrong or significantly exaggerated. Ioannidis believes that the false claims are due to a number of factors including poor research designs, poor statistical reasoning, and the tendency to reinforce the prevailing bias.

Lehrer's (2010) essay expands on Ioannidis' thesis. Lehrer notes that the strength of certain results weakens as time goes by. A group of antipsychotic drugs, for example, had been tested on a large group of schizophrenic patients and were found to be very effective in reducing the patients' psychiatric symptoms, but subsequent research showed that the effect of the drugs was waning. In fact, over time, the effect has been found to be half that of the original. Lehrer cites similar findings of diminishing effects of a variety of variables in a variety of scientific disciplines.

Particularly troubling in the behavioral sciences is the problem of definition. Greenberg (2011) reported on the difficulties faced by the psychiatrists in charge of revising the The *Diagnostic and Statistical Manual of Mental Disease* (DSM-IV). Taking the term, "autism," for instance, Greenberg shows how the definition of this psychiatric diagnostic label has changed across the editions of the DSM. The editor of the present DSM, Allen Frances, told Greenberg that there is just no definition of a mental disorder, and "these concepts are virtually impossible to define precisely with bright lines at the boundaries." Frances' own solution to the definition problem was to use the checklist method, "If a person showed seven of eleven symptoms of a psychiatric disorder, then that person was diagnosed with having that disorder." This definitional strategy produced reliability in that a person diagnosed independently by two psychiatrists usually received the same diagnoses. The problem with this strategy is that it is just a grouping of symptoms. It reflects little understanding of the pathology being diagnosed.

Definitions can have serious consequences particularly when they come from a prestigious group like the American Psychiatric Association. Frances told Greenberg that serious mistakes that had terrible consequences were made in the DSM-IV. For example, diagnoses of autism, attention-deficit hyperactivity disorder, and bipolar disorder skyrocketed. Frances believes that the DSM-IV facilitated these epidemics, and in the process fostered an increasing tendency to chalk up life's difficulties to mental illness and then treat them with psychiatric drugs. Definition issues also cloud the research literature. If, for example, the definition of autism in the twenty-first century is different from what it was in the twentieth century then it is misleading to compare results found from one century to another.

In medical practice it has been noted that today's surgery is tomorrow's malpractice. Surgeons have studied brains and have treated brain pathologies using the best practices of the time. But those same surgeons recognize that the procedures they use today will be obsolete tomorrow. For example, frontal lobotomies were performed by striking the patient with an ice pick in the orbital cavity right above the eye. This today would be considered malpractice. Early surgical treatments for Parkinson's disease involved rather hit or miss invasions of the brain. And, while no present-day treatment for Parkinson's disease is efficacious for all patients, the use of the drug, L-dopa, and deep brain electrical stimulation methods are successful with some patients and are more humane than earlier surgeries. Surgeries are still performed for severe cases of epilepsy, but this is no longer routine as drugs are more frequently used. Some film goers could even support this point retrospectively. In the movie, *Star Trek III*, the starship crew discovers a human skull from way back in the 1990s that shows evidence of a craniotomy. The crew is astonished that people at one time were so barbaric as to have cut into the human skull and into the brain. In short, scientific progress and patient treatment progress are often yoked.

Scientific progress is driven by metaphor. This is the belief of Jonathan Miller (1978) as expressed in his book, *The Body in Question*. Miller suggested that progress in medical science requires the appropriate metaphor. For example, learning about the heart and circulation was facilitated by knowledge of the water pump and the distributive tributaries of the pump's output. In like manner, learning about the nervous system was facilitated by knowledge of the telephone switchboard and of electrical circuits, and most recently, of course, there have been innumerable brain-computer interfaces (BCIs). Like all metaphors there is never complete agreement between the two entities being compared. For instance, neural transmission is not purely electrical; instead it is electro-chemical.

Science can also progress through borrowing. This thesis is put forward by Murray (2009) in his book titled *Borrowing Brilliance.* Although Murray's book deals primarily with innovation in business, he nonetheless notes that innovation in all fields is driven by innovators' borrowing ideas from others. He quotes Einstein as saying, "The secret to creativity is knowing how to hide your sources." And only after Isaac Newton was accused of stealing in the creation of calculus did he say his famous line, "Yes, in order to see further, I have stood on the shoulders of giants." Murray notes that since ideas are born of other ideas, a fine line is drawn between theft and originality. One source from which scientists have borrowed immensely is nature. Ideas pilfered from nature trigger no patent infringement suits. Mosquitoes were using their hypodermic needles long, long before health care professionals; walking sticks were using the power advantage of levers long before physicists began studying levers; snakes' ability to home in on the exhalations of warm-blooded animals preceded the development of heat-seeking missiles; the bat's sonar was in operation long before it was developed by navy scientists; and the Wright brothers watched the birds on the outer banks of the United States' east coast prior to their first flight. The list could go on and on. Suffice it to say that nature is one of the best innovators.

Scientific progress is technology driven. With the invention of the microscope by van Leeuwenhoek in the eighteenth century, what was previously invisible was now visible. The microscope allowed Louis Pasteur in the nineteenth century to validate germ theory. This marked the beginning of modern medicine. The scanning techniques discussed in chapter 1 and in more detail below are prime examples of technology driving scientific progress. fMRI in particular has allowed brain researchers to go beyond restricting themselves to subjects who are first patients presenting themselves to be treated for brain pathologies. Now subjects can be used in behavioral, non-invasive experiments free of any radiation. Computers, too, have advanced scientific progress in neuroscience. First of all, they are very much involved in creating the images that are constructed from brain scans. Second, they can be used to compensate for (not treat) deficits in brain function. BCIs or brain-computer interfaces are in beta testing in a number of centers. BCIs, for example, can be used to help paralyzed individuals control and manipulate the environment. They, moreover, can be used to help non-speakers speak. Locked-in brain individuals have used their brain waves to communicate by sending messages to a computer to indicate which letters, words, or phrases should be selected to have the computer's voice chip compose and synthesize a spoken message.

The detection and rectification of errors also contributes to progress in scientific research. As Nobel Laureate Georg von Bekesy (1960, p. 3) said:

> One of the most important features of scientific research is the detection and rectification of errors. The writer believes that positive results and failures ought to be discussed together. Only by such complete reporting can we get a true conception of a piece of work, of the manner of its development, and of the limitations of its principles.
>
> One way of discovering errors is to repeat the same measurements by different methods. If the same results are obtained by widely differing methods we can feel reasonably confident of their reliability. A way of avoiding errors is to work in a team. The several members can supplement one another's skills and check the procedures. In team research, however, it often happens that the dull, routine work is left to the younger members and not checked by the more experienced. If, for example, the least able individuals are given all the calibrations to do, their mistakes can affect all the results.
>
> Another way of dealing with errors is to have friends who are willing to spend the time necessary to carry out a critical examination of the experimental design beforehand and the results after the experiments have been completed. An even better way is to have an enemy. An enemy is willing to devote a vast amount of time and brain power to ferreting out errors both large and small, and this without any compensation. The trouble is that really capable enemies are scarce; most of them are only ordinary. Another trouble with enemies is that they sometimes develop into friends and lose a good deal of their zeal. It was in this way that the writer lost his three best enemies.

## Issues in Neuroscience Research

Certain research issues keep occurring in neuroscience research. Included are:

## Causality v. Correlation

Probably no student in an introductory statistics course has ever escaped having the teacher bring up some rather silly correlations: a correlation between women's skirt lengths and the Dow Jones Industrial Average or the height at which an insect builds its nest in the summer in Maine and the number of inches of snow in that same state the following winter. These examples are given to introduce the notion that correlation does not equal causality. In brain research, for example, researchers may find a high correlation between a concentration of oxygenated blood in a certain area of the brain and subjects' scores on an introversion/extroversion scale, but does the introversion/extroversion score cause the blood concentration? The question that the researchers should always ask is, "What else could it be?"

## Localization v. Mass Action

The work of Karl Wernicke and Paul Broca (referenced in chapter 2) suggests that brain functions are localized. Wernicke found that a certain area in the left temporal-parietal area of the brain was responsible for interpreting incoming language messages. Broca found that a certain area in the left temporal area of the brain was responsible for encoding linguistic expressive units. If either one of these areas was damaged, certain predicted dysfunctions were manifest. Truly these physician-researchers would support the assertion that brain functions are localized. Karl Lashley, on the other hand, would not. In his research with rodents, Lashley found that large portions of the brain could be removed surgically with no long term deficit.

## Time v. Space

An electroencephalograph (EEG) shows almost instantly that there is a change in brain activity; it does not show where, however. Functional magnetic resonance imaging (fMRI) shows the location of brain activity but only after one to five seconds. Each method has a strength, and each method has a weakness. Some researchers see this situation as being similar to Heisenberg's Uncertainty Principle. Proposed in 1927 (Bronowski 1973), Heisenberg held that one could identify the location of an electron but then be unable to determine the electron's speed and direction, and conversely, if one could identify the speed and direction of the electron, then one could not identify its location. Regardless of what was done there was always uncertainty. The Uncertainty Principle may no longer apply to the EEG-fMRI, time-space issue because recently researchers have simultaneously recorded subjects' EEGs and fMRI responses in experiments.

## Reasoning v. Emotion

The ancient Greeks talked about *logos* and *pathos*. These terms were particularly important in the practice of rhetoric. Does the speaker use a logical appeal (*logos*) to persuade an audience or does the speaker use an emotional appeal (*pathos*)? Freud's concepts of *ego* and *id* are cognates of logos and pathos. From the Greeks to modern Freudian psychoanalysis, logic and emotions became separated. More recently, however, Damasio (1994) found that logic and emotion are linked, and that if there is damage to the emotional centers of the brain, the sufferer cannot make a decision. Emotion is necessary to make a logical decision.

## Automatic v. Thought Driven Behavior

Recalling the Gazzaniga quote at the beginning of chapter 2, the brain seems to do quite well all by itself. At a very basic level we don't have to think about digestion, breathing, heart beat, blood flow, and a whole host of biological processes. At a more advanced level people do not have to look at and think about every step as they ascend or descend a staircase. Reed and brass musicians can often send messages to fellow musicians as they play their instruments. The songs have been played so many times that their playing has become automated. Much of our behavior is automated, although the brain is very active. With the development of fMRI early behavioral studies were event related. For example, if we showed contrasting pictures to a subject while the subject was being scanned, what were the brain's reactions to the contrasting pairs? More recently, researchers have been asking, "What is the brain doing when it is doing nothing?" The answer to this question is identical to the double negative used to describe the continuity of nonverbal communication: "nonverbally nothing never happens." It is the same with the brain: "nothing never happens."

## Neurogenesis v. No Neurogenesis

As noted above, for years neuroscientists believed that we never create new neurons; that the neurons we lose each day as we age are never replaced. Recent research, however, is challenging this view. But the issue is still in dispute, some researchers counter by saying that what appears to be neurogenesis, the formation of new neurons, is really just the build up of new synaptic connections.

## Plasticity v. Rigidity

This issue parallels somewhat the mass action-localization issue. To what extent does the brain change and can one part of the brain dedicated to one task take over the functions of another part of the brain? The fact that people learn and can remember what they learn suggests that the brain is plastic, that it does change. Moreover, the fact that portions of the human brain can be removed, as has been done in cases of severe epilepsy, and the patient post-recovery can still regain pre-surgical cognitive and linguistic functioning supports the plasticity side of the issue. But the brain does seem to become more rigid with age. For instance, Yost's Law, "first learned, last lost," suggests that plasticity may be attenuated with age.

## Monism v. Dualism

This issue can also be restated as the mind-body problem. Moreover, the issue can also be stated thus: "Is the mind merely a function of the brain or is the mind a separate but closely related element?" Early on, strict dualists like Rene Descartes, saw the mind as a non-material element and the brain as physically constituted. More recently, the term, mind, has become almost a synonym for consciousness. Thus, sensation and motor behaviors would be functions of the brain while memory, feelings, reasoning, consciousness, and consciousness of being conscious would be functions of the mind. In this newer view, the mind is not independent of the body. Penfield (1975) has said that the mind has a being distinct from the body, although it is intimately related and dependent on the body.

# Methods in Neuroscience

## The Study of Pathologies

Early studies of the brain were a byproduct of treating patients suffering from a variety of physical pathologies or psychopathologies. In a *physical pathology* there is known tissue or structural damage. *Psychopathologies* are characterized by a variety of dysfunctions that exist in the absence of known tissue or structure damage. However, much research today on psychopathologies focuses on finding physical causes for psychopathologies. For example, schizophrenia, at one time, was seen to be caused primarily by a dysfunctional family situation. Now, however, that view is challenged and schizophrenia is seen to be, in part, a neurotransmitter problem.

Common, non-mutually exclusive categories of physical pathology are autoimmune, congenital, degenerative, infectious, neoplastic, traumatic, and toxic.

An *autoimmune pathology* is one where it seems that the body is attacking itself. An example would be multiple sclerosis. In multiple sclerosis the motor nerves become scarred, interfering with messages getting to the muscles, and as a result, disturbing the motor behavior of the sufferer. Multiple sclerosis or MS can become progressive, and the sufferer becomes progressively more disabled, or there can be periods of remission, in some cases, for extended periods of time.

A *congenital pathology* is one that is acquired in utero or at birth. A common example is cerebral palsy. In the vast majority of cerebral palsy cases the brain is damaged at birth and this damage can be manifest in mental retardation, ataxia, spasticity, sensation and perception problems, and speech and language problems. The neuromuscular status of the cerebral-palsied individual remains static throughout life. Treatment consists primarily of teaching compensatory strategies.

Alzheimer's is an example of a *degenerative physical pathology* that affects the brain and that eventually ends in dementia. A degenerative pathology worsens throughout life. The pace of the degeneration may be slowed or halted temporarily through treatment, but there is no cure. Again treatment consists primarily of teaching compensatory strategies, although drug therapies are now being tried to slow or halt the progression of the disease.

Encephalitis is an example of an *infectious brain pathology*.

Brain cancer is an example of a *neoplastic pathology*, one characterized by the growth of a tumor or new tissue. A recent victim of this pathology, of course, was U.S. Senator Ted Kennedy.

*Brain traumas* can be endogenous or exogenous. A common endogenous brain pathology is a stroke. Strokes can be hemorrhagic or ischemic, the former being the bursting of a blood vessel and the latter the blockage of a blood vessel in the brain. The effects of ischemic strokes can often be reversed if treated early using a clot busting drug called tPA. Exogenous brain pathologies are externally caused. Included here would be the closed head injuries caused by blunt force trauma. Soldiers who are in the field of an explosion often show no initial signs of damage but subsequently will show mild to severe neurological deficits. Some of these closed head injuries are so severe as to cause death.

Heavy metals (e.g., cadmium, lead, and mercury) are *toxic* to the brain. In days gone by, when all well-dressed men wore hats, hat cleaning was a thriving business. The cleaners cleaned the hat and then blocked it with the heavy metal mercury. Many of these workers became mad because of the effects of mercury, thus the term, "mad hatter."

Four common psychopathologies believed to be to some extent brain-based are *autism, bi-polar disorder, depression, and schizophrenia*. The pathologies are not believed to be wholly brain based since some genetic and environmental factors seem to contribute to the symptoms of the disorders. Often, to test for predispositions, researchers will look at families where there are adoptions. Let us say schizophrenic parents adopt a child whose biological parents are not schizophrenic. This scenario tests the effects of environment and contrasts with the scenario where non-schizophrenic parents adopt a child whose biological parents are schizophrenic. This scenario tests the effects of genetic influence. When studies such as these are conducted with psychopathologies, the results are generally interactive in that there are genetic predispositions for certain psychopathologies, but there are environmental influences as well. The brain becomes a factor in understanding psychopathologies since drugs are often used for symptom management. When drug treatment is successful, inferences about the role of the brain can be made.

## Recording the Electrical Activity in the Brain

About twenty years ago in the neuroscience laboratory of Giacomo Rizzolatti in Parma, Italy, there was a lull in the investigatory activity. Sitting passively was a monkey that had electrodes implanted in its brain. The electrodes were also attached to a computer. When one of Rizzollati's colleagues reached for something, there was a burst of activity from the computer that was connected to the monkey's brain (Iacoboni, 2008). This was not simply an example of research recording electrical activity inside the brain, but it also serendipitously led to the discovery of mirror neurons. The mirror neuron topic will be discussed extensively in chapter 6.

## Electrical Stimulation to the Brain

In mid-twentieth century, landmark research was being conducted in the surgical theaters at the Montreal Neurological Institute. Operating on patients to relieve the symptoms of severe epilepsy, Penfield and his colleagues (Penfield and Roberts 1966) electrically stimulated the exposed brain (the brain itself feels no pain). This methodology produced many findings that advanced the field of neuroscience. More recently, researchers have begun exploring the effects of deep brain stimulation to relieve the symptoms of Parkinsonism, Tourette's Syndrome, and depression.

## Scanning Techniques

### *CAT (Computerized Axial Tomography)*

Developed in the 1970s, CAT scans combine many two-dimensional x-ray images to generate cross-sections of three-dimensional images of internal organs and body structures, including the brain. Doing a CAT scan involves putting the subject into a special, donut-shaped x-ray machine that moves around the person and takes many x-rays. Then, a computer combines the two-dimensional x-ray images to make the cross-sections or three-dimensional images. Cat scans, of course, involve radiation, and some subjects must ingest a radioactive substance before the scan.

### *EEG: Electroencephalograph*

It is believed that the electrical activity of the brain was first recorded by Hans Berger in the 1920s. The EEG is not technically a brain scan as the term is used today, but it is, nonetheless, a useful way of non-invasively observing human brain activity. An EEG records electrical signals through electrodes placed on a subject's scalp. The electrodes pick up the electrical signals produced by the brain. These signals are then transduced into mechanical energy, permitting the intensity and frequency patterns of the signals to be ink-traced on continuously moving recording paper. The EEG can show split-second changes in brain activity and thus can show how long it takes before the brain begins to process various stimuli. The EEG is also "portable" in the sense that helmet-like devices can be constructed, permitting subject mobility. A drawback of the EEG, however, is that it cannot identify the anatomical location of an activity in the brain.

### *MRI: Magnetic Resonance Imaging*

MRI is a non-invasive, non-radioactive technique that permits the examination of soft tissues of the body. At the heart of the technique is the fact that, when cells are bombarded by radio waves, hydrogen waves are knocked out of line. As the atoms move back to their position of rest, faint electrical signals are emitted. The MRI scanner uses its very strong magnets to detect these signals. A computer then uses these signals to create a detailed image of the body's soft tissues.

### *DTI: Diffusion Tensor Imaging*

DTI is important in examining the white matter in the brain. The technique is based on water diffusion along the axons. The technique cannot only give information on the integrity of the axons, but can also indicate the direction of axonal impulses.

### *fMRI: Functional Magnetic Resonance Imaging*

When nerve cells are active they consume oxygen and glucose. The response to this neural activity is increased blood flow to the region of the increased neural activity. The increased blood flow to the area of increased neural activity occurs in about one to five seconds and lasts about four to five seconds. The relationship between brain activity and blood flow was first observed by Roy and Sherrington in the 1890s, their observations providing the foundation for the development of modern fMRI technology. Research done approximately one hundred years later by Ogawa and Kwong led to today's fMRI scanner. fMRI measures the hemodynamic response to neural activity in the brain. It is noninvasive and does not involve the radiation inherent in other scanning methods (e.g., CAT scans). It can record a spatial resolution in the region of three to six millimeters but, compared to EEG, has poor temporal resolution, on the order of seconds. fMRI subjects must lie essentially motionless in a scanner during the scan. Because fMRI is noninvasive and involves no radiation, subjects can be scanned for sustained periods, up to two hours in a single session, and for repeated sessions thereafter. Typically subjects may be asked to view films, listen to sounds, smell odors, perform cognitive tasks, press buttons, make moral or ethical decisions in response to problems presented to them, and so on.

### *MEG: Magnetoencephalography*

MEG measures the very faint magnetic fields that emanate from the head as a result of brain activity. MEG uses magnetic detection coils bathed in liquid helium, which cools the coils to superconducting temperatures. The coils are placed over the subject's head. The brain's magnetic field induces a current in the coils which in turn induces a magnetic field in an extremely sensitive instrument called a superconducting quantum interference device or SQUID. At present a MEG device costs millions of dollars and weighs about eight tons. Measuring activity down to milliseconds, MEG provides the most accurate and precise resolution of the timing of nerve cell activity when compared to other brain scanning methods.

### *PET (Positron Emission Tomography) Scans*

Developed in the 1970s, PET scans facilitate the observation of blood flow in the brain. In a PET scan, the subject is injected with a small quantity of radioactive glucose. The PET then scans the absorption of the radioactivity from outside the scalp. Since brain cells use glucose as fuel, PET works on the assumption that if brain cells are more active, they will consume more of the radioactive glucose than other, less active parts of the brain. PET scanners produce color-coded brain maps and they can "observe" deep brain structures.

### *TMS (Transcranial Magnetic Stimulation)*

Transcranial magnetic stimulation works by creating a transient magnetic field under a copper coil placed above a subject's head. The magnetic field induces an electric current in the brain region under the coil. This is called a TMS pulse. Rapid TMS pulses interrupt brain activity. For example, TMS in Broca's area disrupts speech. TMS is similar to Penfield's use of sodium amytal to disrupt certain brain functions including speech.

### *Optogenetics*

Using optogenetic technology, neuroscientists can control neurons in freely moving animals. Saey (2010), in a feature article, has outlined this new technology for studying the brain. Basic to this technology is making neurons sensitive to light. This is brought about by inserting created light sensitive molecules into neurons, or by borrowing such molecules from microorganisms and inserting these borrowed molecules.

When the neurons are made light sensitive the neurons can be "turned on" or "turned off." When a mouse, for example, can optically control the stimulation of the reward system in its brain by pressing a bar, it will repeatedly do so. On the other hand, when the mouse has the opportunity to optically stimulate the fear circuit in its brain, it will freeze after only one pairing. The optogenetic technique is "the new kid on the block" and has been used up to this date primarily with fruit flies and mice. What recommends its continued development is that it is not only an observation method but an activation and suppression method as well. By selecting specific neurons to activate researchers can control the input to an organism as well as the response to the input.

## Summary

The development of methodologies over the years has enabled brain researchers to widen the focus of their research. Research findings were once the by product of medical treatment. Research results were available post-mortem. Then more sophisticated methods were developed but many of these methods involved radiation which necessarily reduced the frequency and duration of scanning. Now brain scans can be done where no radiation is involved. This means that the research subject pool has been vastly expanded. No longer must a subject be a person carrying a patient diagnosis. The only limitation is the tolerance of the subject for the research protocol. In the future we can only expect that brain research technologies will become less intrusive enabling research subjects to generate brain activity data in a variety of unencumbered environments.

# *Chapter 4*
# Human Communication

## Categorizing Human Communication

One method of categorizing human communication is the dimension of *number*. With this method the following forms of human communication can be realized.

- Intrapersonal Communication
- Interpersonal Communication
- Small Group and Team Communication
- Public Address
- Mass Media Communication

*Intrapersonal Communication* is communication with one's self. It is the form of communication that we engage in when we think, problem solve, worry, remember, dream, daydream, fantasize, and so on. In intrapersonal communication we are both the sender and receiver of the message. Much of our intrapersonal communication involves thoughts and fantasies about others, that is, much of our intrapersonal communication is dedicated to the interpersonal. Awake or asleep the intrapersonal communication "machine" keeps churning. It simply cannot be stopped. The importance of intrapersonal communication cannot be overestimated for it is this form of communication that gives us the opportunity to plan, to resolve problems, to think about self and others, and to escape through fantasy if the demands of life become too burdensome. All you need to do is just close your eyes and you can be anywhere you want to be and be with anyone you want to be with. Even dreams, the ore mined by the psychoanalysts, should not be discredited. The famous film director, Stephen Spielberg, once said, "I dream for a living" and "I interpret my dreams one way and make a movie out of them and people see my movies and make them part of their dreams" (see brainyquotes.com). Spielberg was talking about his sleeping dreams and his daydreams, both good examples of intrapersonal communication. It does not seem to be an accident, therefore, that the company he co-founded is called, *DreamWorks*.

We engage in intrapersonal communication when it appears that we are doing nothing. But, we remind ourselves that in the brain nothing never happens. Even when we appear to be engaged with another person we can be miles away thinking about or daydreaming about other people and other places. We never know what is going on inside another person's head even when that person seems to be listening and giving the requisite eye contact and head nods.

A great deal of neuroscience research attempts to obtain a threshold answer to what is going on inside another person's head. Stimuli are presented through various sense modalities, and scans are performed to see which areas of the brain seem to be activated. Do the stimuli seem to trigger approach, pleasure, avoidance, distress, cognition, and so on?

*Interpersonal Communication* refers to communication between two individuals. It occurs primarily in the face-to-face setting but can of course be in written and electronic modes as well. It is through interpersonal communication that the notion of self and other emerges. Throughout life there are significant others who are instrumental in the acquisition and maintenance of our self-concepts. The psychoanalyst, Erik Erikson (1963), believed that throughout life there is always a significant other. For the first year of life it is the mother; for the second year it is the father. Early development, according to Erikson, was, therefore, interpersonal.

*Small Group and Team Communication* involves three or more participants. The actual upper limit on the number of participants varies but there is some agreement that the number of participants should be small enough so participants can establish eye contact with every other participant, and be able to hear other members. Egolf and Chester (2007a) put the upper limit at nine, while other writers have gone as high as twenty to twenty-five. In Erikson's developmental scheme significant others after age two become collective. Parents of adolescents know that the peer group becomes most important for their children. Later the work group is instrumental in the maintenance of self-concept. A dramatic change occurs when the number of participants goes from two to three, that is, when interpersonal communication changes to small group communication. With three or more participants, coalitions can form. Thus, two participants in a three-member group can form a coalition against the third member. This cannot happen in the interpersonal setting. Coalition forming, coalition break-ups, and the emergence of new coalitions constitute one of the key dynamics of small group communication. If you ever experienced living in or working in a three-person situation, you may have experienced the two-against-one phenomenon. Two children form a coalition against a third, two roommates form a coalition against the third, two workers form a coalition against the third, and so on.

Another key dynamic of small group communication is directly related to number. As the number of participants in the group increases, the opportunity for any one member to talk lessens. For example, with three people in a group, in an hour's time, each person would be able to talk for twenty minutes if talking time were shared equally, but with each additional member added to the group, the time each member has to speak diminishes. Going from three to ten members, for instance, reduces each member's talking time from twenty to six minutes. As numbers increase, therefore, members spend more and more of their time listening.

In *Public Address Communication* there is usually one speaker addressing an audience. Thus, all but one of the participants serves in a passive role. Interaction is very limited, limited to a few questions or comments. Common public address situations are classroom lectures, banquets, political rallies, and so on. Again, the laws of physics are in play. As the number of participants increases, the opportunity for interaction decreases.

*Mass Media Communication* is characterized by a single *agency* sending a *mediated message* through a given *medium* to a *mass audience*. The sending agency is some government organization, C-Span, for example; a nonprofit agency, PBS, for example; or a commercial one, ABC, CBS, FOX, and NBC, for example. A mediated message is one that has been modified in some way so that its relationship to an actual event may be remote or detached. This modification is triggered by the medium used and is not necessarily the result of deliberate bias. For example, regardless of how hard they try to be objective, a TV crew sending a live feed of a demonstration back to the studio nonetheless modifies or alters the situation by the choices they make in directing their cameras, in the selection of interviewees, in the time they stay at the demonstration, etc.

Even in reality TV, there is no reality because any communication situation fed through a technological filter or through a technological medium is changed. Common media, of course, are print (newspapers and magazines, for example) and electronic (radio, computer downstreams, blogs, television, cell phones, and email, for example). The mass audience for mass communication messages can indeed be massive, often measured in the millions for popular television shows.

## A Human Communication Theory

Harper (1979) has developed a theory of communication that can be diachronically or synchronically applied. Diachronically Harper applies the theory from the classical period in ancient Greece through contemporary times. Diachronically and synchronically, the theory can be applied to any of the contexts described above whether the message is verbal or nonverbal (see below). There are five components to Harper's theory:

- *Categorization or Memory*: This component refers to the communicator perceiving events and storing a record of these events so that the events can be recalled,
- *Conceptualization or Invention*: this component refers to the process of converting information into knowledge through the process of interpretation.
- *Symbolization or Style*: This component refers to the process of meaning making through the process of turning knowledge into symbols.
- *Organization or Arrangement*: This component refers to the ordering of symbols and noting the relationships between and among them.
- *Operationalization or Delivery*: This component refers to the transmission of information and knowledge through a given medium. The selection of a medium and the manipulation of media is included in this category.

To analyze communication Harper notes that if one is analyzing the sender, one begins analyzing at Stage 1, categorization or memory, and proceeds stepwise from there through Stage 5. On the other hand, if one is analyzing the receiver of a message, the analysis would begin at Stage 5 and then work backward from there through the steps.

# Categories of Human Communication Messages: Verbal Communication and Nonverbal Communication

## Verbal Communication

Verbal Communication is communication through the use of language. A *language* is a system of symbols known by at least two individuals. The key words in this definition of language, as noted, are *system* and *symbol.* When language is referred to as a system, it means that there is a set of rules in any given natural language (for instance, English, German, Japanese, Chinese) for putting the structural units of the language together in order to produce an utterance. In English, for example, "Man bites dog" is quite different from "Dog bites man," suggesting, of course, that there are word-order rules in English. In Latin, for example, word order is less important.

A *symbol* is a *sign* that stands in an *arbitrary* relationship with a *referent.* A sign is something that stands for something else, the referent. Essentially all words are symbols; that is, they stand for or represent something and are therefore first of all, signs. But since words stand for or represent things in an arbitrary way, since they are not like what they stand for or represent, they are also symbolic. We call the thing we sit on "chair," but there is no inherent connection between the thing, chair, and the word, "chair." This is demonstrated by the fact that members of other language communities call the thing people sit on by other names. In other words, meanings generally are what any particular language community decides the meanings should be. A few words are not symbolic, however. Words like "buzz," "pop," and "thud," for instance, sound much like what they represent, the buzzing of bees, the pop of a champagne cork, the dull sound of an object falling. These words are not arbitrarily related to their referents. They are like their referents but not in an arbitrary way, and, therefore, are signs but not symbols.

It is this arbitrariness of most words that makes language so flexible and gives us so much leeway and ease in terms of what we can speak and write about. Think of all the concepts, abstractions and formless, invisible entities that language enables us to deal with by labeling them with words. Verbal communication messages can be spoken, written, or signed as in American Sign Language (ASL). Linguists agree that ASL, satisfying the system and symbol requirements, meets all the requirements of a language.

## Aspects of Language

### *Language as a Model of the Universe*

With a PhD in chemical engineering in hand, Benjamin Lee Whorf (See Carroll 1964) became a fire inspector for a large insurance company. Whorf noted that workers would not smoke around drums marked "Full" but would smoke around drums marked "Empty." Of course, this was a tragic mistake for the "empty" drums were far more dangerous since gasoline fumes are far more explosive than liquid gasoline. This observation of the "empty-full" behavior aroused Whorf's interest in language and led him to study language and culture, at times collaborating with Edward Sapir, from which emerged The Whorf-Sapir Theory of Linguistic Relativity. A brief description of this theory follows.

Whorf recognized that there are an infinite number of stimuli in the world but that our brain, in spite of its one hundred billion neurons, has only a finite capacity. This means that we are hopelessly overwhelmed; there is simply too much to deal with. One solution is to categorize the stimuli, that is, we call some things chairs, and ignore the differences among individual chairs. In this way, we do not overload our brain by trying to find a new name for every chair in the world (for, in fact, no two chairs are exactly alike).

Thus, *categorization* of stimuli simplifies our life, and we categorize by using language. Whorf found that people use language to categorize or simplify the world. Different groups categorize in different ways based upon what is important for the group. For example, for Trobriand Islanders yams are the staff of life, and, as a result, an elaborate yam vocabulary developed; for Eskimos and some skiers snow is very important and led to the creation of an elaborate snow vocabulary. For the Westerner a yam is just something bought at the market and the person simply cannot "see" or "know" about the yam as does the Trobriand Islander. They live in two different worlds.

Whorf was fond of saying that your language is your map of the territory. What he meant by that is that there is a physical world that is potentially open for everyone to know, but there is too much to know so through language we categorize, paying most attention to the things important to us and ignoring things less important. Our language, in short, is our map of the territory. But Whorf reminds us that, just as there can be many maps for any territory, street maps, road maps, topographical maps, demographic maps, subterranean maps, there are many language maps, English, French, German, Japanese, Russian and Spanish, for instance. No map covers all the territory and, by not doing so, it gives its user only one view of the world.

Whorf believed that the language you speak determines the way you think and perceive so all your thoughts and perceptions are relative to your language. Other language users think and perceive in different ways. These different ways of thinking and perceiving or seeing the world are most contrasted between different languages. Whorf contrasted the languages of native North Americans, the Hopi, for example, with the languages of their European conquerors, noting how members of the two groups could be in the same physical environment but truly living in two different worlds. At the same time, even within the same language culture, we can see some Whorfian effects. For example, members of any specialty, be it medicine or mechanics, learn elaborate vocabularies to direct their perception and thought so that they literally live in a world different from the world of lay people; the specialists and lay people can be looking at the same human body or the same automobile, but the specialists see things to which lay people are "blind."

## *General Semantics*

*Semantics* refers to the study of meaning, and *General Semantics* is a branch of semantics which focuses on the effects of words and labels on people's perceptions, attitudes, feelings, and behaviors. Words and labels can affect the way we perceive ourselves and others; they can affect our attitudes or our predispositions to believe and act; they can determine how we feel, triggering any one of a range of emotions; and they can affect our behavior, mobilizing us to go in one direction or another. Many people have contributed to the General Semantics literature; we can only sample from this vast literature to support the general semanticists' position. The following general semantic examples testify to the power of words and labels on people.

The Self-fulfilling Prophecy: Rosenthal (Tauber 1997) completed studies in which elementary school children were given an IQ test with an impressive name, "The Harvard Test of Inflected Acquisition." The actual scores of these tests, however, were not given to the teachers. Instead, the teachers were told that certain students were expected to bloom and were labeled "bloomers." The teachers accepted these labels and believed them, assuming that the children so labeled were quite capable of learning. The teachers taught them accordingly. The children did indeed learn, in fact, beyond the expectations that their actual test scores suggested.

The net effect of labeling the children as "bloomers" affected the teachers' perceptions and attitudes toward the children and mobilized the teachers to behave in a manner that assumed the children could and would learn. This effect is known as the *self-fulfilling prophecy*, and Tauber says, "*What the self-fulfilling prophecy process does is label someone and then have that person treated as if that label were correct.*" And earlier, according to Tauber, Merton described the concept as follows: "*The self-fulfilling prophecy is, in the beginning, a false definition of the situation evoking a new behavior which makes the originally false conception come true.*" The whole process begins most often with words and, thereby, testifies to the power of words.

The Placebo Effect: When a new drug is being tested, the clinical trials include giving some patients the real pill and others a placebo or sugar pill, encapsulated to look like the real pill. The drug and the placebo are administered in a double blind manner, meaning that neither the dispenser of the pill nor the patient knows which pill is the real pill and which is the placebo. Pill dispensers are not told and are kept "blind" to whether they are dispensing the real drug or the placebo in order that they do not verbally or nonverbally transmit the message that this will or will not help. Similarly, patients are kept "blind" to what they are taking so that they maintain the belief that what they are taking will have therapeutic value. Why is this done? The simple answer is that if one is given a pill (even a placebo) and led to believe that it has therapeutic value, it often does have therapeutic effects. Somehow some healing process is triggered by the label, "pill," and the physical reality of the pill. Frank (1974) listed a number of pathologies that respond positively to placebos alone. Included in Frank's list are both physical pathologies and psychopathologies.

Other General Semantic effects can be found in the following anecdotes.

- A four-and-a-half-year-old girl goes to the dentist. At the end of the session the dentist says, "You did so well today I thought you were at least five years old." The little girl is on cloud nine, talking about it on the way home and rushing to tell her father. The five-year-old label elevated her self-esteem.
- Clerks in retail stores are no longer clerks but associates, a label that is supposed to make them part of the team with decision-making power.
- Weapons of mass destruction are called "Peacemakers" in the hope of changing attitudes about the weapons.
- Virtually all weapons are called defensive weapons, suggesting that no war will ever be initiated. How could it? There are no offensive weapons.
- And the label, "War Department," has long ago been changed to "Department of Defense."
- Tax increases are called "revenue enhancements" in the hope that voters will react less angrily to the higher taxes.

- Patients are often labeled according to their prognoses, "terminal," for example, or according to diagnosis, "There's a kidney in 404."
- Hospital floors are often referred to by patient prognosis; "That's terminal up on ten."

The last two anecdotes suggest how labels may affect the way staff members perceive and treat patients, often in negative and dehumanizing ways.

## *Metaphor*

A *metaphor* is a very powerful figure of speech. When we use a metaphor, we say something is analogous to something else, and we make this comparison in the hope that what we know about the latter can be applied to the former in order to help us understand the former. As noted above, in his book, *The Body in Question*, Miller (1978) stated his belief that metaphors were very powerful tools in understanding the human body. We didn't know much about the heart until the water pump was invented and little about the brain and nervous system until switchboards and computers were developed. Aiding medical researchers, therefore, were the metaphors: the heart is a pump, the nerves are electrical lines, and the brain is a computer. The philosopher, Francis Bacon, had a preference for insect metaphors. He compared philosophers in the Aristotelian tradition to spiders, always weaving a web, enticing, and ensnaring their prey in little word games that never lead to practical knowledge. On the other hand, he compared the experimental method practitioners to bees, animals that go out and gather the pollen (data), process it (data analysis), and then turn it into honey (results that have practical use). In recent years, the spider's web metaphor, of course, has been most prominent with the internet. Developed as a failsafe system that could remain operational even if part of it suffered damage, the internet has all the characteristics of a web with a worldwide span of threads and nodes.

Other functions of metaphors are:

- Metaphors make discussion more vivid and colorful. A person who is a bit unstable, for example, can be described as unstable, or can be described more vividly as a "loose cannon," a "quart low," a "less than full deck," a gas stove whose "pilot light has gone out," and so on. While the metaphor may not always be fair, it can liven up the discussion nonetheless.
- Metaphors can facilitate the remembering process. This may be due to the fact that metaphors create pictures in the mind, and visual images can contribute to the remembering of associated linguistic information. Also involved is the emotional factor. Metaphors often trigger an emotional reaction, either positive or negative, and emotion and memory are yoked together. Think of any traumatic event that you experienced and you will probably remember the minute details of anything you saw, heard, or did on that day.
- Metaphors can motivate. The teachers in Rosenthal's self-fulfilling prophecy study were influenced by the flower metaphor. The students were late bloomers. The flower metaphor can extend in the mind. Late bloomers

must still be watered and cared for in order for the late blooming to be realized. In like manner late-blooming students must be taught with the belief that they too will bloom, albeit late.

### *Vagueness and Ambiguity*

Words can be *vague*, that is, their meaning can vary from one person to another, or their meaning lies on a continuum. Vague words are not categorical. "Warm" lies on a continuum and does not have a single meaning. For a given person a day may seem to be warm but to another person, that same day may seem to be cool.

Words that can have multiple meanings are called *ambiguous*. Advertisers and politicians routinely use ambiguity in there ads. Why? Because ambiguous messages allow receivers to supply their own meanings to the ambiguous messages. Ambiguous messages are like the figures in the Rorschach test. Viewers of the items supply the meaning. A political ad which says, "Jones is the man," says nothing; it lets the recipients of the message supply the meaning.

### *Denotative and Connotative Meanings*

Definitions of words that we find in the dictionary are *denotative*. The definitions represent what most people would agree are indeed the meanings of the words. Frequency of usage is key here. *Connotative* definitions, on the other hand, are peculiar to the individual. Often connotative definitions are formed as a result of an intense emotional experience. For example, a therapist is testing the speech and language skills of an elderly woman who has had a stroke. Using standard procedures the therapist is using a "floor-ceiling approach." Finding a floor means the therapist will find a block of test items that the patient will answer correctly. The therapist will then administer progressively more difficult items until the patient can no longer answer the items correctly. This is the ceiling. The floor items on the test call for simple imitation. The therapist says, "Buy some bread," and the patient is asked to say the same words. All is going well until the therapist says, "Sell the house." Immediately the patient says, "No!" and begins to cry uncontrollably. The therapist is puzzled and seeks to find an answer from other staff members. It turns out that on the very same day of testing, the patient's children, who had gotten power of attorney over their mother, had sold the mother's house, the house where she lived her entire adult life. "House" had acquired a very tragic connotative definition for the patient.

### *Words Can Reveal*

Psychologists, psychiatrists, and psychoanalysts often listen to what a patient says and then ignore the threshold meaning. Instead, these practitioners accept another meaning that they believe reveals the content of the patient's unconscious. For instance, a psychiatrist listens to an elderly patient talk repeatedly about how poorly nursing home patients are treated. Is the patient talking about a social problem? No. The patient is really expressing a fear of having to go to a nursing home and being poorly treated. Slips of the tongue are another way words can reveal. A key source listing a number of slips of the tongue or Freudian slips can be found in Freud's *The Psychopathology of Everyday Life* (most recently re-published in 2009). What most people hear as a simple slip of the tongue, Freud sees as the unconscious coming to the surface. One simple example reported by Freud occurred when Freud was seeing a female patient, and at the end of the session Freud said, "Good day, sir." Freud then realized that he suspected that the woman's husband was eavesdropping at the door and that throughout the session he was addressing the husband and not the patient.

## Nonverbal Communication

By *nonverbal communication* is meant that form of communication wherein messages are sent by virtue of an agent's internal bodily activity (including brain activity), physical characteristics, adornment, touching behavior, body movements and postures, facial expressions, eye behavior, utilization of time, vocal behavior, use of space, objects, odors, and taste. Unlike language, there is no recognized system for combining or connecting nonverbal units. Therefore, nonverbal communication lacks a system or set of rules for combining expression units in spite of the attempts of some researchers to find such a system.

Nonverbal units, however, may be both sign and symbol; that is, there are nonverbal units that may be just signs, but nonverbal units may also be symbolic by virtue of the fact that they are arbitrarily related to a referent or have the potential to be symbolic. A red traffic light or a model or picture of a traffic light, for instance, is like a red traffic light and might be used, as sign, to represent a red traffic light; but symbolically or arbitrarily, it means stop. And objects or models or pictures of grain or rice can stand, as sign, for grain or rice, but they may be thrown at a wedding because they symbolize fertility; they are not thrown to represent grain or rice.

### *Categories of Nonverbal Communication*

The field of nonverbal communication is traditionally broken down into a number of categories. These categories are sometimes referred to as *codes* or *modalities*. Whatever the designation, categories, codes, or modalities, the purpose of the sub-categorization of nonverbal communication is to point out and emphasize the different ways people communicate nonverbally. It is these categories around which many nonverbal textbooks are organized. Each category becomes the central theme of a textbook chapter in which that category is discussed in detail. Defined below are eleven categories of nonverbal communication.

### *Vitalics*

*Vitalics* refers to the effects of the body's vital signs on communication. Included here are skin conductance, muscle tone, heart rate, blood pressure, respiration rate, and brain activity, for example. These vital signs, of course, communicate to the physician, who monitors these message systems for signs of physical pathology, hypertension, for example. For the nonverbal observer, however, these message systems can provide information against which verbal messages can be compared. One example of such use is polygraphy. Here the polygraph operator (the polygrapher) asks questions and listens to the subject's answers to those questions while simultaneously observing the subject's vital signs, skin resistance, blood pressure, and respiration rate, for example. If vital signs showing marked changes accompany the answers, the polygrapher suspects that the subject is lying. In neuroscience the neuroscientist looks at brain activity for messages that reveal the state of a research subject. For instance, fMRI can reveal that a subject is stressed, fearful, or afraid if the amygdala "lights up." Physicians look at MRIs of the brain for evidence of tissue damage

### *Organismics*

*Organismics* refers to the effects of physical characteristics on communication. Even when still, our bodies send a multitude of messages by virtue of their size, shape, height, weight, physical attractiveness, and so on. While we like to think that "it's what's inside that counts" or that "you can't tell a book by its cover," research indicates otherwise. As soon as we present ourselves, decisions are made about us, our intelligence, personalities, social skills, and so on. So "looks" do indeed communicate.

### *Cosmetics*

*Cosmetics* refers to the effects on communication of changing the appearance of the body through invasive and non-invasive procedures. In present times, the main reason for using these procedures is to make ourselves more attractive. This testifies to the effects of physical characteristics, particularly attractiveness, discussed above. Since the beginning of recorded history people have tried to make themselves more attractive, many times at great risk. Lead-based makeup certainly made the skin the sought-after white but at a high price, since this heavy metal caused brain deterioration. Today, of course, we cannot watch television without seeing shows on cosmetic surgery. More and more people are seeking the services of cosmetic surgeons, not necessarily for vanity reasons, but because they know that bodies communicate, and certain bodies communicate more positive messages than others.

### *Costuming*

*Costuming* refers to the effects of dress on communication. Dress is a primary modality for expression. The number of different messages we can communicate via dress is enormous. Each day we make decisions about how we should dress. With each selection, we are also selecting a message to send to ourselves and to others. Even if our dress follows the line of a Kris Kristoferson song, and we put on our "cleanest dirty shirt," we have still made a decision about communication. Just as physical appearance sends messages to others before we say a word to them, so does our dress communicate. In many cases, nonverbal communication begins long before the verbal begins.

## Haptics

*Haptics* refers to the effects of touch on communication. Touch is the most basic form of communication; it is the form that we first experience when we make our entries into this life, and it is most likely, if we are lucky, the form that we will experience when we make our final exits. When words fail, people touch. They touch in times of joy, in times of sorrow and in times of hate or anger, for instance. They do the high fives, they embrace, and they strike out. Truly, touch is a modality for sending emotional messages. Touch is also important in maturation, in healing (therapeutic touch), in communicating status, and in communicating cultural ties, for example.

## Kinesics

*Kinesics* refers to the effects of body movements and postures on communication. This area of nonverbal communication is sometimes referred to as "body language," a term which has become popular in non-academic circles but which is technically incorrect since nonverbal is not a language (see definitions above). Body postures and movements communicate extensively, revealing, for example, if people are relaxed or apprehensive, competing or cooperating, male or female, or want to hold the floor or relinquish it when speaking. In fact, according to one researcher, body postures may even communicate something about peoples' personalities during sleep.

## Personics

*Personics* refers to the effects of facial expressions on communication. The face is the most expressive part of the human body. It is the part of the body to which interactants pay the most attention, and it is the part of the body that most identifies an individual. When people are asked to think of someone, they most frequently picture that person's face. *Persona* refers to a mask, and we refer to facial communication as personics because the face is capable of making so many masks, so many distinct expressions. It is as if we all have a plethora of masks that we can change at a moment's notice in order to communicate a plethora of messages. How many? Estimates vary. Some estimate that the face can communicate five thousand distinct expressions. That estimate is based on the possible permutations of movements of all parts of the face. In reality, we do not discriminate among that many expressions but, instead, focus on far fewer expressions.

## Oculesics

*Oculesics* refers to the effects of the eyes on communication. The power of these two small orbs in the head is amazing. Have you ever changed your behavior just because someone was looking at you? Eyes can invite, mesmerize, intimidate or censure, show likes and dislikes, and apprehension and attentiveness, for instance.

### *Vocalics*

*Vocalics* refers to the effects of the voice on communication. A speaker says words, but the way in which the words are spoken constitutes the nonverbal or vocalics part of the spoken word. The same words can be said loudly or softly, rapidly or slowly, dysfluently or fluently, with an accent or without an accent, for instance. In each case the vocalic part of the verbal message acts as an editorial with the potential of changing the meaning of the spoken message. Take the very simple declarative statement, "John is a doctor." By changing stress and timing patterns you can speak this statement in ways so that one utterance of the statement will contradict the meaning expressed in a second utterance of the statement. The voice is a very sensitive communicator of emotion. When someone calls you on the phone, you can tell within seconds how the caller feels.

### *Chronemics*

*Chronemics* is the effects of time on communication. Time can be a key communicator of status. Those of high status wait less whether it is for a table in a restaurant, an answer to a phone inquiry, or for service in any number of establishments. Time can show liking and preference for people and activities. Observe how individuals spend their time, and you know whom and what they like. Time can communicate emotion. When people go from glad to sad, the tempo slows.

### *Proxemics*

*Proxemics* refers to the effects of space on communication. Two categories of space are of interest. The first, *dynamic* space, refers to the ever-changing aspects of space among people. Among strangers, for instance, people are always dividing up the available space. The first occupants to enter an elevator position themselves at the corners, and then look at the floor numbers. Occupants who enter later take positions that divide the remaining available space. What happens when a large number of people exit a crowded elevator simultaneously? They immediately spread out, once again dividing up the now much larger available space. The second category of proxemics, *static* space, refers to the relatively fixed features of space. Architecture and design, how buildings are constructed and arranged in relation to each other, how cities are laid out, for example, fall within the proxemic category. Static space is a key communicator of status. Those of high status have more space, higher space, and more private space, for instance. The study of proxemics also involves the concept of *territory*. We all establish territories. Fathers and mothers have their places at their tables at home, students have their seats in the classroom, and residents of city streets claim certain parking places as their own even though they are public spaces. All organisms, in fact, display proxemic behaviors.

The placement of objects in spaces is not surprisingly called *objectics.* Many writers include this category as part of proxemics. People communicate by collecting objects that enhance their images, automobiles, houses, furniture and so on. Architects, designers and decorators thrive on these inclinations. In like manner people use scents and tastes to create more appealing spaces. Perfumes and colognes are designed to make the user more appealing and acceptable to others. Controversy swirls around the extent to which people bond or are attracted to one another because of olfactory stimuli. Advertisements for pheromones (substances that send covert olfactory messages) claim that their products will attract the opposite sex. There seems to be little support for these claims. At the same time there seems to be some support emerging for the hormone oxytocin's (often called the love hormone) having a very positive effect among people who inhale it. Serving tasty food not only satisfies basic nutritional needs but its "yummy" smell can communicate something positive about the server.

## Comparing and Contrasting Verbal and Nonverbal Communication

Below are a number of factors on which we can compare and contrast verbal and nonverbal communication. The factors are presented to sharpen the distinction between the two.

### *Primitiveness*

Nonverbal communication is more primitive both *phylogenetically* and *ontogenetically.* The root of "primitive" is *primo,* meaning first. So the statement indicates that nonverbal communication came first; it preceded language in two realms. The first realm is the phylogenetic realm. *Phylogenetic* means evolutionary development. If we look at the developmental span of the species in this universe, we note that the first form of communication was nonverbal. To this day, the vast majority of messages in our universe are nonverbal. The lowly insects, for example, communicate using *pheromones,* scents which carry messages. The hated cockroach leaves a slime trail, providing much information for other roaches about the sender and its itinerary. Ants that work cooperatively and fight cooperatively also communicate via pheromones. Such communication is nonverbal, and it works quite well. We can traverse the evolutionary ladder and find elaborate nonverbal patterns of communication in animals. A bee's dance communicates about food supply; a bird's song sends mating, protection, and migration messages; dogs and other animals mark territory, and chimps and dolphins, who are said to have some linguistic skill, communicate virtually exclusively through the nonverbal medium.

*Ontogenetic* refers to development from birth onward. No one who is reading this book was speaking words on the day of birth. You probably did not speak even single words until you were between seven and twelve months of age. In fact, infants communicate not only at birth, but even while still in the womb. At birth, infant and mother communicate through touch, movement, and sound. Prior to birth, infants send messages through movement, heartbeat, and images on sonograms, for instance. And it has been reported that in utero the unborn child responds to sound. So in the developmental life of the infant, it is the nonverbal that comes first. The verbal comes later, two-word combinations appearing at around eighteen to twenty-four months of age.

### *Continuity*

Verbal communication is discontinuous. People stop talking, they stop writing, and stop signing. Nonverbal communication, on the other hand, is continuous; we cannot stop it. This continuity factor has led to a number of double-negative statements that describe this aspect of nonverbal communication: "Nonverbally you cannot not communicate." And, "Nonverbally nothing never happens."

Goffman (1963) not only referred to the continuity factor in his statement, but also alluded to its social hazards: "Although an individual can stop talking, he cannot stop communicating through body idiom; he must say either the right thing or the wrong thing. He cannot say nothing."

### *Analog/Digital*

Nonverbal communication is said to be *analog* communication while verbal communication is said to be *digital*. An *analog* message is one where the message and the meaning merge. Most people wear analog watches, for example. An analog watch is an analog of time. We think of time as something going around; first it was believed that the sun was going around the earth, but later it was found that the earth was not only rotating, but also moving around the sun. At any rate, there is rotation that defines time, and the hands on an analog watch rotate, preserving this sense of time. A digital watch, however, gives no indication of this rotation. The message and the meaning do not converge. The message is remote from the meaning in a digital message. In fact, when we look at many high-tech digital watches we do not know if we are reading the time, a stop watch number, or the wearer's blood pressure, for instance.

A person may say, "I'm mad at you." This is a verbal or digital message. It can be quite remote from the meaning of anger. But, if the speaker shouts the words, if his face reddens, he bares his teeth, flares his nostrils, and clenches his fists, we have a good indication that the speaker is, indeed, angry. These non-word messages, of course, are nonverbal, and they are analog messages because they begin to merge or become equivalent to the definition of anger.

### *Tense*

Nonverbal communication is primarily present tense communication. It communicates about the here and now. Verbal communication, on the other hand, can communicate about the past, the present, and the future. These tenses exist in spoken, written, and certain sign languages (ASL, for example). If we try to create a sense of time in nonverbal communication, we often use fragments of very rudimentary sign languages. For example, when people who are unfamiliar with any sign language are asked to signal past tense nonverbally, they point backward; in like manner, when asked to indicate the future, they point forward. Interestingly, these are the signs used in a rudimentary sign language called AMERIND (American Indian Sign Language). AMERIND was a language used by Native Americans to communicate with the white settlers who came to their lands and with other Native American tribes who spoke different languages.

### *Emotional/Informational*

Nonverbal communication is the chief transmitter of emotion, while verbal communication is the chief carrier of information. Listen to any newscast describing a tragedy. At times the newscaster may be calm and collected, just reporting the event in words. At other times, however, the death, dying, suffering, and devastation witnessed and being reported are overwhelming, and the newscaster "breaks down." The newscaster's speech may become dysfluent, and the newscaster may even cry. It is these breakdowns that communicate emotion, emotion communicated nonverbally.

### *Referencing the Negative*

Verbally we can designate things that are not present and may not exist, that have never existed, and may never exist. We can easily talk about the absence of real or imagined people, or even pink rhinos, unicorns, wizards, and flying elephants. How much more difficult it is, though, to refer to or reference such absent entities nonverbally. How, for instance, might you inform a group of listeners that there are no flying pink elephants in the room? Can you do that more efficiently verbally or nonverbally? Referencing the negative, that is, referring to the absence of something, is accomplished with significantly greater ease through the verbal channel, thus reinforcing the notion that nonverbal communication deals primarily with the here and now.

### *Neurological Correlates*

It is probably more than coincidence that the division made in communication between verbal and nonverbal parallels is the division found in the human brain. Psychobiologist Sperry (1975), a pioneer in hemispheric specialization, said:

> Now both the left and right hemispheres of the brain have been found to have their own specialized forms of intellect. The left is highly verbal and mathematical, performing with analytic, symbolic, computer-like, sequential logic. The

> right, by contrast, is spatial and mute, performing with a synthetic spatio-perceptual and mechanical kind of performance processing that cannot yet be simulated by computers.

Thus, the left side of the brain is dominant for speech and language reception, processing, and expression, and for analytic thought. The right side of the brain, on the other hand, is dominant for nonverbal processing, the recognition of faces, the recognition of time, the processing of spatial relationships, and the recognition of melodies, for instance.

This neurological division was discovered by first observing the changed communication behaviors of individuals who had survived severe head trauma, and later upon their deaths, performing autopsies to pinpoint the location of brain damage. Communication behaviors and sites of damage were then examined and correlations between various behavior patterns and sites of damage identified. It was found that left brain injuries tended to lead to speech and language dysfunction, from rather mild word-finding difficulties to full-blown disabilities involving the reception, understanding, and expression of speech and language. Right brain injuries tended to lead to nonverbal communication dysfunctions. For example, a right-brain-injured person might have severe problems in getting the right shoe on the right foot and the left shoe on the left foot because the part of the brain that handles spatial relationships has been injured.

Later support for the hemispheric division of labor came from Sperry's split-brain experiments in which the fibers connecting the two brain hemispheres were severed; and from experiments in which radioactive nutrients or gases are administered, and the brain is scanned while subjects complete a variety of tasks. For example, when research subjects drink a radioactive sugar solution or inhale a radioactive gas and then have their brains scanned while they listen to music, the right side of the brain is active, but, when subjects are asked to analyze the music, the left side of the brain becomes active.

Sperry (1975) noted that the brain's division of labor has implications for education:

> A message that emerges from the findings of hemispheric specialization is that our educational system and modern society generally (with its very heavy emphasis on communication and early training in the three Rs) discriminates against one whole half of the brain. I refer, of course, to the non-verbal hemisphere, which we find, has its own perceptual, mechanical, and spatial mode of apprehension and reasoning. In our present school system, the attention given to the nonverbal hemisphere of the brain is minimal compared with the training lavished on the left hemisphere.

Goleman (2006, 2007), for example, has tried to end the discrimination against the right hemisphere in his books, *Emotional Intelligence* and *Social Intelligence.* The basic thesis of Goleman's two books is that a large component of success in life is dependent upon right hemispheric functions.

### *Metacommunication*

The literal meaning of "meta" is "beyond," so the word "metacommunication" carries some sense of beyond. More commonly, *metacommunication* is defined as communication about communication. Nonverbal communication serves a metacommunicative role by virtue of the fact that it often communicates about a speaker's verbal message. It is a kind of editorial about the verbal message. The shivering person who denies being cold is sending a metamessage. The nonverbal metamessage of shivering "says," "I am cold," and this message contradicts the verbal message of "I am not cold." It is the metamessage that the skilled communicator seeks and often uses to test the validity or truth of the verbal message. When the words say one thing and the body "says" another thing, which do we believe? Most skilled communicators believe the body.

### *Complementary/Contradictory*

Given the discussion of metacommunication above, it follows that when verbal and nonverbal messages occur simultaneously (as they must do in human communication), the two messages stand in one of two possible relationships to one another: *complementary* or *contradictory*. *Complementary* means that two things go together or support each other. Therefore, complementary messages would include the following examples:

- One person says to another, "Good to see you," and shakes hands warmly and heartily.
- A mother says to her child, "Come to mommy," and kneels and holds out her arms to receive the child.
- A diner says, "The food was great," and eats with gusto.

In addition to the shivering example given above, other examples of contradictory messages are:

- An interviewer says to an interviewee, "I have all the time in the world," but repeatedly looks at her watch.
- A person shouts, "No, I am not defensive!"
- After hearing someone tell a joke, a listener says, "That was funny," without vocal emotion and while displaying a facial expression of sadness.

## Defining Communication

There are many definitions of communication. In fact, Dance and Larson (1976) presented 126 published definitions of communication. The large number testifies to the complexity of the phenomenon. The rather lengthy discussion of verbal and nonverbal behavior above was intentional. It was designed to testify to the complexity of human communication. Even when two people are trying to achieve complete understanding between them, communication is difficult. People speak in metaphors, they use ambiguous or vague words, they have connotative meanings for words not shared by the other, and they have had different life experiences which places them in different worlds from the outset. Then add to this the fact that people are also deceptive. In some cases the deceptions are benign. People say they feel great when they do not. They compliment their interaction partners when the compliment is not meant. Some deceptions, to be sure, are more gravitas. A spouse lies about an affair, a parent lies to a child, a job applicant lies on a resume. All the deceptions in life complicate the communication process. Is it any wonder that there are so many attempts at defining communication?

We will not attempt to list even a small sample of the definitions of communication on the Dance and Larson list. Instead, the list of definitions below is designed to show the path that leads to fidelity (the sense of the word used here comes from acoustics where the input to a system matches the output) between sender and receiver. This is Definition No. 9. Fidelity between sender and receiver is rare. This is what makes communication so frustrating and, at the same time, so exhilarating.

1. Communication is.
2. Stimulus-response connections are communications.
3. All behavior is communication.
4. Communication is behavior exhibited by a sender that is consciously noticed by a receiver.
5. Communication is behavior exhibited by a sender that is consciously noticed by a receiver and to which the receiver ascribes meaning.
6. Communication is meaningful behavior exhibited by a sender without any conscious notice given to it by a receiver.
7. Communication is meaningful behavior intentionally exhibited by a sender that is consciously noticed by a receiver.
8. Communication is meaningful behavior intentionally exhibited by a sender that is consciously noticed by a receiver and to which a receiver ascribes meaning.
9. Communication is meaningful behavior intentionally exhibited by a sender that is noticed by a receiver and to which the receiver ascribes meaning, a meaning identical to that intended by the sender.

As you look at the nine definitions you will see a progression along a dimension that ranges from the very general to the very specific. Definition 1 is the most general and Definition 9 is the most specific. Key factors leading to the increasing specificity are *intent, meaning,* and *consciousness.* Definition 1 is the most general, suggesting that as long as an individual is in a state of being or conscious, that person is communicating, communicating to self or to others. This definition applies particularly to brain activity. Recall that in the brain, "Nothing never happens." When we are doing nothing the brain is active, even in sleep. We dream when sleeping and daydream during our waking states. Our corporeal presence, whether we are awake or sleeping, communicates to others.

Definition 2 holds that stimulus-response bonds are communication. An individual may or may not be aware of stimulus-response messages. Actually unawareness in most cases is a blessing. Imagine being aware of every neuron-to-neuron transmission. The result would be maddening. Less maddening but severely disturbing would be an awareness of the stimuli that trigger each heart beat, the secretion of digestive juices, and the feedback that adjusts the body before, during, and after each walking step. A number of studies reviewed in this book will be of the stimulus-response type. For example, a subject in an fMRI scanner is shown a photograph. What response does the photograph (the stimulus) trigger in the brain of the subject? A response is suggested from the scan with the subject possibly being unaware of the process.

Adopting the definition that all behavior is communication does not assume that the sender intentionally sent the message or that the receiver received it. It is possible that the message was intended and that the message was received, but intent is not a requirement for this definition. This definition would accept the notion that when two people are communicating, one or both of the participants can be aware or unaware that communication is taking place. In short, communication can occur above and below the level of awareness. Definition 3 is an "anything goes definition." But as we proceed from Definition 3 onward, certain requirements are increasingly added before we can call behavior communication, until we reach Definition 9, which has the most stringent requirements.

Which definition is the correct one? They all can be. It really depends upon the situation and the need. The medical diagnostician, the psychiatrist, the clinical psychologist, the spy, and the detective assume that all behavior is communication, and they try to find the meaning of even the most minute and seemingly insignificant behaviors. Much of the time, of course, meanings cannot be found and ascribed. Thus, those who adopt Definition 3 do, indeed, work in a clouded semantic environment. But the assumption is made that, potentially at least, all behavior is communication. In some situations this cloudiness cannot be tolerated. The ground crew worker who hand signals a plane to the gate, for example, must send a message whose intended meaning is the identical meaning ascribed to the message by the pilot. In this situation, Definition 9 must be in force. If the congruence between the ground crew and the pilot did not exist, the likelihood of an accident would dramatically increase.

## Summary

The upshot of all the above is that the studies reviewed in this book will sample only a small part of the communication universe. This is due to both the complexity of the brain and the complexity of communication. What remains is still endlessly fascinating and marks the progress that has been and will be made in the field.

*Part II*

# Introduction

Part 2 examines the role of the brain in various communication contexts (intrapersonal, interpersonal, and mass) and then examines communication disorders that can affect individuals' ability to communicate. Finally, ways to ameliorate the effects of communication disorders, primarily through the use of technology, are detailed.

# *Chapter 5*
# Intrapersonal Communication

In the previous chapter intrapersonal communication was defined. Briefly it is communication with one's self; it is sometimes called self-communication. One person is both the sender and receiver of a message. Major categories of intrapersonal communication are a present state of being, sleeping and dreaming, default situations, thinking, problem solving, memory, and consciousness. Of all the communication contexts in which individuals participate, the most time is spent in intrapersonal communication. Even when we are talking to another person our minds can be wandering, we can be daydreaming, fantasizing, problem solving, worrying, and so on.

## Communicating a Present State of Being

The normal brain is always communicating signals about its current activity state. These electrical signals can be captured using electroencephalography, commonly referred to as EEG signals. Electroencephalography is non-invasive; signals are captured with pickups attached to the outer skull. The EEG of an individual is described in terms of the amplitude and rhythmic activity of these electrical signals across time. When portrayed on an oscilloscope or a paper tracing, the electrical activity generates sine-like waves, and cycles per second can therefore be recorded as a result. In general, several categories of brain wave activity have emerged, each communicating some present activity state in the brain. The categories are:

*Beta* (12 to 30 cycles per second). Beta is the brain wave state communicating concentration, alertness, thinking, anxiety, worrying, and problem solving, for example.

*Alpha* (8 to 12 cycles per second). Alpha has been called the brain wave state of meditation. Indeed meditation leaders have their students enter the alpha state by repeatedly reciting a mantra or through biofeedback procedures. An Indian Swami once described the alpha state as nowhere. People often go into this state without the benefit of meditation, watching TV, sitting in front of a fireplace, watching the ocean at the beach, or just sitting at one's desk, for instance.

*Theta* (4 to 8). Entering the theta state one becomes drowsy, and theta can communicate that one is entering an early stage of sleep.

*Delta* (0 to 4 cycles per second). An individual in delta is in deep sleep. Of course, virtually everyone has heard the term "flatliner" which refers to a patient who continuously exhibits a flat or zero-cycles-per-second EEG for a sustained period of time. This most often refers to a brain dead diagnosis. The whole question of at what point do you communicate that you are dead is at the present time being debated in the literature and will be discussed below.

EEG signals can be transmitted to a physician, a medical technician, a researcher or to one's self, as might happen in a biofeedback situation. In addition for those individuals with severe motor problems, voluntarily generated EEG patterns can be used in brain-computer interfaces.

## Sleeping and Dreaming

Sleep is often described as occurring in four stages. As the sleeper proceeds from Stage 1 to Stage 4, more and more time is spent in the Delta brain-wave state. Van De Castle (1994) has described the four stages:

- **Stage 1.** A Stage 1 sleep pattern is characterized by a mixed-frequency EEG pattern with a concentration of activity in the 2 to 7 cycles per second range. Stage 1 sleep is light sleep. It is a state between being awake and falling asleep. There are no sleep spindles, K-complexes, or delta waves in Stage 1 sleep. A sleep spindle refers to electrical activity of 12 to 14 cycles per second having the same amplitude as theta waves. The spindle name was motivated by the sleep spindles' resemblance to the spindle on a loom. A K-complex is a single high amplitude wave that stands out from the background theta waves. It stands out because it is so much higher than the background theta waves. Delta waves are the low frequency waves described above.
- **Stage 2.** Stage 2 sleep is described by the presence of sleep spindles and K-complexes. Delta waves in Stage 2 sleep make up less than 20 percent of the EEG recording. Stage 2 signals the real beginning of sleep. The sleeper becomes disengaged from the environment. Body temperature drops. Approximately 50 percent of a night's sleep for a normal adult is spent in Stage 2 sleep.
- **Stage 3.** Stage 3 sleep contains between 20 and 50 percent delta waves. Sleep spindles and K-complexes are also present. In Stage 3 breathing also

becomes slower, blood pressure drops, muscles relax, and the body is restored. Approximately seven percent of a night's sleep for a normal adult is spent in Stage 3 sleep.

- **Stage 4.** When more than 50 percent of the EEG activity consists of delta wave activity, the sleeper is in Stage 4. This is the defining characteristic of Stage 4. Otherwise, Stage 4 sleep is similar to Stage 3. Approximately 14 percent of a night's sleep for a normal adult is spent in Stage 4 sleep.
- **REM (Rapid Eye Movement) Sleep.** REM sleep occurs in conjunction with Stage 1 EEG pattern. The difference between Stage 1 non-REM (NREM) and REM sleep is that in REM sleep there is the presence of rapid eye movements which are absent in NREM sleep. For the normal adult REM sleep occurs approximately 22 percent of the time during the night and NREM sleep occurs seven percent. During NREM sleep the sleeper's heartbeat is fairly steady, but during REM sleep, the sleeper's heartbeat is irregular, alternating between rapid and slow heartbeats. In like manner, a sleeper's breathing in NREM sleep is slow and regular, but in REM sleep, the respiration rates vary from very fast to very slow.

Below is a paraphrase of Van De Castle (1994, p. 233) summary of sleepers' nights:

> As sleepers move from wakefulness through drowsiness they pass through a very brief descending sleep from Stage 1 down to Stage 4. After having been asleep for an hour or so they move from Stage 4 back up through Stage 3, and eventually Stage 1 sleep reappears, though it may only last four or five minutes. Following this brief plateau, which may or may not be accompanied by REM sleep, the sleeper usually returns to Stage 4 sleep. Approximately 90 minutes after the onset of the first Stage 1 plateau the sleepers tend to display REM sleep which may last up to 10 minutes. Subsequently sleepers will generally return to Stage 3. REM sleep will usually return in approximately 90 minutes and the cycle repeats itself.

# What Do Dreams Communicate?

## Dreams Can Forecast the Future

From the beginning of time dreams have been seen as predictors of the future. For example, the book of Genesis, chapter 41, tells the story of Joseph interpreting the dreams of the Pharoh:

> PHARAOH dreamed: and behold, he stood by the river. And behold there came up out of the river seven well-favored kine, and fat-fleshed, and they fed in a meadow.
>
> And behold, seven other kine came up after them out of the river, ill-favored and lean-fleshed, and stood by the other kine, upon the brink of the

> river. And the ill-favored and lean-fleshed kine did eat up the seven well-favored and fat kine: so Pharaoh awoke. And he slept and dreamed the second time; and behold seven ears of corn came upon one stalk, rank and good. And behold, seven slim ears and blasted with the east wind, sprang up after them. And the seven thin ears devoured the seven rank and full ears: and Pharaoh awoke, and behold, it was a dream. And it came to pass in the morning that his spirit was troubled, and he sent and called for all the Magicians of Egypt and all the wise men thereof: and Pharaoh told them his dreams: but there was none that could interpret them unto Pharaoh. Then the chief butler of Pharaoh made known unto him the skill of Joseph in the interpretation of dreams, and Joseph was brought out of the prison into which he had been cast by Potiphar, his master, and Pharaoh related unto him the dream which had perplexed him. And Joseph said unto Pharaoh, The dream of Pharaoh is one; God hath shewed Pharaoh what he is about to do.
>
> The seven good kine are seven years, and the seven good ears are seven years: the dream is one. And the seven thin and ill-favored kine that came up after them are seven years; and the seven empty ears blasted with the east wind shall be seven years of famine. This is the thing which I have spoken unto Pharaoh: what God is about to do, he sheweth unto Pharaoh. Behold, there come seven years of great plenty throughout all the land of Egypt. And there shall arise after them seven years of famine, and all the plenty shall be forgotten in the land of Egypt, and the famine shall consume the land. And the plenty shall not be known in the land, by reason of that famine following, for it shall be very grievous. And for that the dream was doubled unto Pharaoh twice, it is because the thing is established by God: and God will shortly bring it to pass. Now therefore let Pharaoh look out a man discreet and wise, and set him over the land of Egypt. Let Pharaoh do this, and let him appoint officers over the land, and take up the fifth part of the land of Egypt in the seven plenteous years. And let them gather all the food of those good years that come, and lay up corn under the hand of Pharaoh, and let them keep food in the cities. And that food shall be for store to the land, against the seven years of famine! which shall be in the land of Egypt, that the land perish not through the famine. And the thing was good in the eyes of Pharaoh and in the eyes of all his servants. And Pharaoh said unto his servants, Can we find such a one as this is, a man in whom the spirit of God is? And Pharaoh said unto Joseph, Forasmuch as God hath, shewed thee all this, there is none so discreet and wise as thou art: Thou shalt be over my house, and according unto thy word shall all my people be ruled: only in the throne will I be greater than thou. And Pharaoh said unto Joseph, See, I have set thee over all the land of Egypt. And Pharaoh took off his ring, from his hand, and put it upon Joseph's hand, and arrayed him in vestures of fine linen, and put a gold chain about his neck And he made him to ride in the second chariot which he had.

The evidence for dreams predicting the future is sketchy at best. Part of the problem has to do with methodologies. It is the age old question of field studies versus laboratory studies. In field studies there is a loss of control. A dreamer reports a dream that predicted a future event. A not uncommon one is that a dreamer dreamed of tomorrow's winning lottery number. The dreamer played that number and won. On the surface the evidence seems to be convincing. What is missing of course is the knowledge of how many other people dreamed about a number that same night that did not win the next day

One area where dreams seem to be accurate in predicting the future is in the field of medicine. Dreams seem to be able to predict the onset of serious illnesses before a diagnosis is made. This seems to be particularly true for cancer. Van De Castle has said that there are many published examples of dreams preceding the onset. Three examples follow:

- A woman had recurrent nightmares of dogs tearing at her stomach a few months before she was diagnosed with stomach cancer. The woman died three months after the diagnosis.
- A journalist had a dream in which torturers were placing hot coals beneath his chin. He felt the heat start to sear his throat and screamed in desperation. He awoke very disturbed and received a call from his girlfriend who had just had a horrible dream of being with him in a bed filled with blood. In a later dream the journalist dreamed that medicine men were sticking hypodermic needles in his neck. The journalist felt sure there was something wrong in his throat. He, however, had difficulty convincing his physician of this. Eventually he was diagnosed with thyroid cancer.
- A woman with breast cancer dreamed that her head was shaved and that the word 'cancer' was written on her head. She awoke with the feeling that her cancer had metastasized to her brain even though there were no physical signs present. Three weeks later her dream diagnosis was confirmed.

What is remarkable about the "cancer diagnosis dreams" is that they are not only confirmed by the subsequent diagnosis, but they pinpoint the site of the tumor as well. The mechanism underlying this process is unknown. Perhaps there is some neural signal from the sufferer's immune system which is becoming compromised.

More recently, Claassen et al. (2010), have collected data suggesting that vivid and violent dreams can foreshadow brain disorders occurring as much as fifty years later. The condition has been labeled REM sleep behavior disorder. The dreams of these individuals often involve fighting off an attacker. The dreamer often goes through the motions of fighting off the attacker, not suffering the paralysis that usually accompanies REM sleep. The neurogenic diseases that were found to develop in these individuals were degenerative including Parkinson's and Lewy body dementia. The disorder seems to afflict primarily males since the twenty-four of the twenty-seven participants who passed the screening for the study were men.

## Dreams and Discovery

Dreams can lead to discovery. One of the most famous examples from the history of science is the dream of Friedrich August Kekule who discovered the structure of benzene after having a vivid dream (Roberts 1989).

> I was sitting writing at my textbook but the work did not progress; my thoughts were elsewhere. I turned my chair to the fire and dozed. Again the atoms were gambolling before my eyes. This time the smaller groups kept modestly in the background. My mental eye, rendered more acute by the repeated visions of the kind, could now distinguish larger structures of manifold confirmation: long rows, sometimes more closely fitted together all twining and twisting in snake like motion. But look! What was that? One of the snakes had seized hold of its own tail, and the form whirled mockingly before my eyes. As if by a flash of lightning I awoke; and this time also I spent the rest of the night in working out the rest of the hypothesis. Let us learn to dream, gentlemen, then perhaps we shall find the truth. But let us beware of publishing our dreams till they have been tested by waking understanding.

A second example from the history of science is that of Otto Loewi who won the Nobel Prize in medicine in 1936 for his work on the chemical transmission of neural impulses.

In 1903, Loewi had the idea that there might be a chemical transmission of the nervous impulse rather than an electrical one, which was the commonly held belief, but he was at a loss on how to prove it. He let the idea slip to the back of his mind until seventeen years later he had the following dream.

> The night before Easter Sunday of that year I awoke, turned on the light, and jotted down a few notes on a tiny slip of paper. Then I fell asleep again. It occurred to me at 6 o'clock in the morning that during the night I had written down something most important, but I was unable to decipher the scrawl. The next night, at 3 o'clock, the idea returned. It was the design of an experiment to determine whether or not the hypothesis of chemical transmission that I had uttered 17 years ago was correct. I got up immediately, went to the laboratory, and performed a single experiment on a frog's heart according to the nocturnal design. (see http://www.brilliantdreams.com/product/famous.dreams.htm)

It took Loewi a decade to carry out a decisive series of tests to satisfy his critics, but ultimately the result of his initial dream-induced experiment became the foundation for the theory of chemical transmission of the nervous impulse and led to a Nobel Prize! Loewi noted: "Most so called 'intuitive' discoveries are such associations made in the subconscious."

## Dreams and Creativity

Certainly there can be a thin line between discovery and creativity. Creativity here will be discussed in terms of artistic endeavors. There are numerous cases where artists have attributed their creative products to dreams. A few will be catalogued below.

Paul McCartney described how the melody for the song, "Yesterday," came to him in a dream in 1965. In McCartney's dream he heard a classical string ensemble playing. McCartney said:

> I woke up with a lovely tune in my head. I thought, 'That's great, I wonder what that is?' There was an upright piano next to me, to the right of the bed by the window. I got out of bed, sat at the piano, found G, found F sharp minor 7th – and that leads you through then to B to E minor, and finally back to E. It all leads forward logically. I liked the melody a lot, but because I'd dreamed it, I couldn't believe I'd written it. I thought, 'No, I've never written anything like this before.' But I had the tune, which was the most magic thing! (see http://www.brilliantdreams.com/product/famous.dreams.htm).

Jimmy Ibbotson (2003), the front man for the Nitty Gritty Dirt Band, told how a song came to him in a dream. Jeff Hanna one of Ibbotson's band members said that one of his favorite gospel songs was Ibbotson's "I find Jesus." Ibbotson replied:

> That's funny. That song came to me in a dream. I was in Costa Rica one night and uh I dreamt I was in this southern church and all the people looked at me and said "get up and sing your song, your new song." Someone tapped me on my back (it was my dad who died months before) and he said, "go ahead Jim, get up and sing your song." The thing just came to me in real time.

Dreams have played a major role in the film world. Below are the thoughts of four famous directors. The quotes from Bergman and Wells were taken from Van De Castle (1994). Material for Fellini and Spielberg can be found on the personal quotes sections of the biographies for the respective directors on http://www.imdb.com/name/nm0000229/bio.

- Film director Ingmar Bergman said, "I discovered that all my pictures were dreams. Of course I understood that some of my films were dreams, but that all my pictures were dreams was a new discovery to me."
- Orson Wells, the American film director was asked about the dreamlike quality of the connected rooms in his film, The Trial. Wells responded by saying, "I attempted to make a picture like a dream I have had. I move from architecture to architecture in my dreams." Asked if the film was about a film character's dream, Wells responded, "No, it's my dream. I dreamed about him."
- Federico Fellini the Italian film director used dreams extensively in his films to produce the many grotesque and surreal characteristics of many of his characters and scenes. Many of Fellini's films dealt with historical themes. Fellini once said that history is science fiction in reverse, and that one could read about ancient Rome but one could know almost nothing about the feelings of the people who lived at that time. One's dreams were as good a source of information as any other.
- Mentioned in the previous chapter was American film director Stephen Spielberg who said "I dream for a living." Spielberg's film, ET, for example, contained a common dream theme, that of the journey of going home. Some psychoanalysts might say that underlying this theme is the search for mother an underlay often said to exist in The Wizard of Oz.

## Dreams Communicate Psychological Dynamics

With the beginning of psychoanalysis, psychiatry, and clinical psychology dreams were the ore to be mined by the healers. Dreams were interpreted by the healers and patients were told what their dreams meant and how they were crucial in diagnosis and treatment. Three representatives of the clinical professions are discussed below.

*Freud*: In 1896 Sigmund Freud (1856–1939), a Viennese neurologist, abandoned neurology and began to analyze dreams, slips of the tongue, childhood memories, and transient forgetfulness. His work was instrumental in producing his book entitled *The Interpretation of Dreams* (Freud 1980) originally published in 1900. Freud concluded that in dreams the unconscious surfaces. Repressed sexual and aggressive desires surface in dreams. Dreams create the royal road to the unconscious. But of course many dreams are highly symbolic and therefore need to be interpreted. Thus the notion of the Freudian symbol emerged. Vertical structures in dreams where symbolic of the phallus while round structures, particularly if there were water involved, a fountain, for example, could be symbolic of the womb.

*Jung:* Carl Jung (1875–1960) was mentored by Freud. Eventually Jung split with his mentor primarily due to their differing views on dreams. Unlike Freud, Jung came to believe that dreams serve to guide your waking life (Jung 1965). Dreams can offer a solution to problems that people face in their waking lives. Imagine that a person has become a doormat in the interpersonal arena. The person receives no respect. This person may dream of being assertive and not being a doormat. Such dreaming may give the dreamer guidelines on how to behave in daily waking situations. In a similar manner, teachers, actors, and others who perform before an audience often dream of facing an audience unprepared; they even may appear before the audience naked. This dream can remind the performer to prepare and mobilize for the performance. It can raise the level of alertness.

*Perls:* Frederick "Fritz" Perls (1893–1970) is considered to be the founder of Gestalt Psychology. In simple language, Perls believed that dreams supply the parts needed to fill life's voids to permit an individual to become a whole person. the quote below (see Perls 1970) shows that Perls clearly supports the notion that dreaming is self or intrapersonal communication:

> The dream is an existential message. It is a message of yourself to yourself. Every part, every situation in the dream, every aspect of it is a part of the dreamer, but a part that to some extent is disowned and projected onto others. If we want to own these parts of ourselves again, we have to use special techniques by which we can reassimilate those experiences (p. 27).

Outside of the clinical transactions themselves two effects emerged. First, because of the sexual connotations attached to many dreams, people became reluctant to talk about their dreams. Dreams reported in the presence of others often trigger derisive laughter, finger pointing, and "pop-psych" responses. Second, many small dream interpretation enterprises emerged. For small sums someone would interpret your dreams. The lines on radio talk shows would light up the board when a dream interpreter would be a guest. Now with the internet an additional outlet exists for dream interpreters to publicize their services. Moreover, books and recordings of various kinds are available for helping people interpret their dreams.

## Why Do We Dream?

The reasons for dreaming are still unknown. Perhaps the brain is just doing a little housekeeping. As with any cleaning project, things get a bit disorganized as some items are put aside for storage while others are discarded. A more scientific argument is that during dreaming the brain stem stirs up strong emotions such as anxiety, euphoria, anger, and fear. Images are then created that portray these emotions. This explanation seems reasonable since most dreams occur during REM sleep when brain stem activity surges. Countering this explanation is one that is in sympathy with Freud. Specifically, this explanation holds that during dreaming the frontal cortex is seeking out objects and images of desire and interest. The debate will continue. At this time it seems that dreams can warn of potential threats, help the dreamer deal with a variety of problems, stimulate creative activity, and diagnose some illnesses.

## The Default Network

Think of all the things that have gone through your mind in the last thirty minutes:

- How many times did your mind drift from the immediate environment and what you were doing in the environment to which you sojourned?
- In your mind drifts did you reflect on the past?
- In your mind drifts did you plan for the future?
- To what locations did you travel in your mind drifts?
- Who were the people involved in your mind drifts?
- Did you fantasize at any point in your mind drifts?
- Why did your mind go drifting?

Mind drifting, mind wandering, or daydreaming is a normal activity and neuroscientists refer to this activity as the brain going into the *default state*. Contributing to the construction of this state is a set of brain areas which in combination are called the default mode network (DMN) named by Raichle and his colleagues in 2001. Areas of the brain generally identified as part of the default network are the medial prefrontal cortex, the posterior cingulate cortex, and the left and right inferior parietal lobes.

Efforts are underway to further discriminate the functions of these areas. For instance, Andrews-Humma et al. (2010) found that the posterior cingulate and the anterior medial prefrontal cortex are active when people make self-relevant, affective decisions. On the other hand, when people are making decisions involving constructing a mental scene based on memory, a medial temporal lobe subsystem becomes engaged. Finally, when people are engaged in future-oriented thought, the two systems are engaged simultaneously. These respective areas become active when the brain is at "rest" or in the default mode.

Raichle (2010) has noted that the evidence suggests that a malfunctioning default network is involved in a number of diseases including Alzheimer's disease, autism, schizophrenia, depression, post-traumatic stress disorder, Tourette's syndrome, amyotrophic lateral sclerosis, and attention-deficit hyperactivity disorder. In general, people suffering from these pathologies show different activity in the default regions of the brain from that found in normal individuals.

Even in conversation participants can go into the default mode when speaking or listening, particularly when "listening." It is sobering to consider that when in the company of others whether it is on a bus, an airplane, or in a theater audience, for example, one has absolutely no idea what is going on in the heads of others. In each brain is a bubbling reservoir of good and evil. People are thinking about the good deeds done for them and the good deeds they will do for others. They are planning for future encounters and thinking of what they can anticipate from others in these encounters. Then, of course, there are the serial killers, the molesters, the sociopaths, and psychopaths feeling their urges and thinking how to satisfy them.

This fact has not escaped theorists who wrote long before modern brain-scan studies. For example, more than fifty years ago Kurt Lewin (Hilgard 1956) noted that we live in both physical space and psychological space. We can be situated in a given physical space but at the same time may be traveling widely in psychological space. Hilgard's (1956), presentation and review of Lewin's theory, provides an example:

> . . . as I sit in reverie and make plans for tomorrow, I move in a world very different from which I sit. My life space is the space in which I live psychologically, as seen from my own viewpoint. It includes each and every object, person, idea, that I have anything to do with. It corresponds in many ways to the world about me, to the world of things, people, and ideas, being my selection from these, but it becomes always my world in edited and distorted forms. It is possible to distinguish between me and my life space, for I move about in my psychological space as others do in theirs. We interact in the real world, all the while producing changes in our respective psychological worlds. (p. 264)

Lewin proposed that when the physical situation becomes benignly (boredom) or severely (being held captive and punished, for instance) intolerable we can

Lewin proposed that when the physical situation becomes benignly (boredom) or severely (being held captive and punished, for instance) intolerable we can always escape into psychological space. This fact has sometimes been used to explain multiple personality disorders. A child who is terribly mistreated may escape in psychological space by taking the identity of another enabling the child to conclude at some level that this isn't happening to me. The normal person moves freely in both spaces. To stay exclusively in one space can be indicative of a psychopathology. Some schizophrenics, for example, retreat from their physical environments and live almost exclusively in psychological space. On the other hand, a person rendered pathologically distractible, who must pay attention to every environmental stimulus, is a prisoner trapped in physical space.

Lewin noted that the life space of an adult is more highly differentiated than that of a child. This raises the question of how the default network develops. Gao et al. (2009) found that a primitive and incomplete default network is present in two-week-olds. As the infant becomes older, the number of default regions exhibiting connectively increases. The researchers found that by two years of age the default network becomes similar to that observed in adults. Brain regions involved in both two year olds and adults are the medial prefrontal cortex, the posterior cingulated cortex, the inferior lobe, the lateral temporal cortex, and the hippocampus.

The emergence of the default network would seem to parallel the emergence of self. What suggests this parallel relationship is that when the brain is in the default state it is often thinking about the past and making and planning strategies for the future, particularly future interpersonal encounters. To function satisfactorily in these endeavors the brain must have developed sufficiently and have had the requisite interpersonal experiences. There are many views of the emergence of self. In the communication field one group that has paid particular attention to the self is the symbolic interaction group of writers and researchers.

Littlejohn (1978) has reviewed the field of Symbolic Interaction and the prominent leaders in the field. A basic premise of Symbolic Interaction is that mind, self, and society are not discrete structures but processes of personal and interpersonal interaction. Self is acquired and maintained through interpersonal communication both at the verbal and nonverbal levels. At every stage in life there are others who are instrumental in shaping the self. These "others" are sometimes called significant or orienting others.

As noted earlier, child development specialist and psychoanalyst Erik Erikson (1963) identified significant others across the span of life. Erikson held that in each of life's developmental stages there are others who have a significant impact on one's concept of self. The eight stages are:

- *Birth through First Year:* The Mother is the Significant Person.
- *Through Second Year*: The Father is the Significant Person.
- *Third Year through Fifth Year:* The Family is the Significant Group.
- *The Sixth Year till the Onset of Puberty:* The Neighbor and School provide Significant-Other Groups.

- *Adolescence:* Peer Groups become significant.
- *Early Adulthood:* Friends, Intimates, and Workmates are the Significant Others.
- *Middle Adulthood:* The Shared Household provides the Significant-Other Core.
- *Late Adulthood:* Family is important but people often see humanity as being significant as they like to come to the conclusion that their lives have made a difference.

Evident in Erikson's model is a significant other or a group of significant others for each developmental stage. In the beginning of life, significant others are the parents and the basic family. Next, the radius of significant others expands to include schoolmates and neighbors; and with adolescence, peer groups take center stage. Later, partners and workmates enter the radius. And, at the end of life, significant others may include relatives, friends, and perhaps personal care workers, but also humanity in general. We want to think about what our life meant and whether we made a difference.

These interactions with significant others across life provide the source for the musings that occur in the default state. These musings can recall the past or think about the future.

Why does the brain go into the default mode? A number of reasons should be suggested by the above paragraphs:

- The mind wanderings may relieve the wanderer from the discomfort of the present.
- The mind wanderings may give an individual the time to simulate future interpersonal encounters.
- The mind wanderings help to maintain a sense of self.
- The mind wanderings may provide the thread that connects the past, present, and future.
- The mind wanderings may maintain a baseline level of arousal in the brain and thus maintain conscious awareness.
- The mind wanderings permit multi-tasking.

## Memory

Memory refers to the ability to capture, store, retain and retrieve information. Sometimes this process is outstandingly simple. Sometimes it is one-trial and permanent learning. Children who have put their tongues on a frosty chain link fence on a sub-zero morning complete the process in one trial, and the lesson remains etched in their brains forever. Contrast this with the same children who are required to know the capitals of all the states in the United States and the process is prolonged and stressed. There are problems of learning, retention, and retrieval. People who have written on the topic of memory generally talk about short term memory (STM) and long term memory (LTM).

## Short Term Memory

You are driving in your car. A phone number for a product or service that interests you is given on your car radio. While you are looking for a safe place to stop and capture the number in your cell phone or to write the number down you keep reciting the number to yourself. As soon as the capturing or writing is complete, you stop the recitations and forget the number but immediately form a new memory: where the number is stored externally. The capacity of our short term memory is limited. Quantitatively we seem to be able to handle five to nine chunks of information. The word "chunks" is used because we can remember information better when it is chunked. 80057462413 is more difficult to remember than the same number when chunked: 800-574-6213. In fact it is difficult for us to recite the previous number without chunking it into the three segments since this is what we do habitually.

Short term memory is very fragile. Imagine a last minute rush to the grocery store. You need bread, milk, and eggs, and you keep repeating those three items to yourself as you rush to the store. But as you enter the store there is a chef at the entrance giving out free samples of a new dessert. You have to taste it. And then the public address system announces that any customer carrying a super shopper card will get the twelve-ounce package of this dessert free if they buy twelve items or more at checkout. The end of this story is many times predictable. You leave the store with twelve items and no bread, milk, or eggs. Distractions can destroy short term memories.

Gazzaniga et al. (2009) present a number of subcategories of short term memory. There is first *sensory memory*. Sensory memory has a lifetime measurable in milliseconds to seconds. For example, when someone is talking to us and we are not paying close attention but are able to recover some of what the person is saying as though we were paying close attention. *Short term memory*, according to Gazzaniga et al., is memory that retains information from seconds to minutes. Remembering a phone number or a person's name would be instances of short term memory. Finally, Gazzaniga et al. talk about *working memory*. Working memory includes the retention of information needed to act on a given problem at the moment. Reviewing a household budget and comparing expenses of one month with a previous month may require the reviewer to hold certain information in memory as the calculations are made.

## Long Term Memory

Some information we can remember for a long time, even for a lifetime. These are long term memories. Common categories of long term memory have been proposed by Carter, et al. (2009) and Gazzaniga, et al. (2009).

Carter and colleagues talk about *episodic, semantic, and procedural long term memories*. Episodic memories involve the reconstruction of past experiences including the sensations and emotions that accompanied and helped define one's experiences. Semantic memory refers to the facts that we commonly call knowledge. A medical student is required to learn the names of the twelve cranial nerves, for example, The student may, in addition, be required to know whether or not the nerves are sensory, motor, or both sensory and motor. Procedural memories are body memories. Everyone has heard that you never forget how to ride a bicycle. Young people who have disappointed their parents by not continuing their piano lessons will long remember and can play a piece that they learned by heart in their early years.

The long term memory categories of Gazzaniga, et al. parallel those of Carter, et al. They begin with the dichotomous categories of *declarative* and *nondeclarative* long term memories. Declarative memories are subdivided into *semantic* and *episodic* memories, the definitions of which mirror those of Carter, et al. Under the nondeclarative category, Gazzaniga et al. have a number of subcategories including procedural memories, perceptions, and simple learned behaviors that result from conditioning.

## *Neurological Correlates of Short and Long Term Memories*

At one time it was believed that there was a single seat of memory in the brain but that position has been virtually abandoned. Memories seem to be distributed in a number of areas in the brain. Evidence for this comes from patients who have suffered brain injury from a variety of brain traumas. For example, a famous patient in the history of neuroscience is Henry G. Molaison (HM). At age twenty-seven, in 1953, HM had surgery (a large portion of the hippocampus was removed) to attenuate the effects of seizures. As a byproduct of the surgery, HM lost his capacity to store any new information in his long term memory. However, he still retained his ability to receive, understand, and produce fluent and grammatically intact speech and language. HM died as a "twenty-seven-year-old" at age eighty-seven since virtually no new information was added to his long term memory reservoir since the operation at age twenty-seven. Evidence also comes from various brain scans. Most would agree that the findings to date are preliminary and more light will be shed on the topic with future research.

With the above caveat in mind some general findings have emerged. According to Carter et al., the seat of short term memories is located in the frontal lobe. Spatial memories seem to be associated with the parietal lobe. The caudate nucleus seems to be associated with memories of instinctual skills. The mammillary body is associated with episodic memories. The putamen is associated with procedural skills. Emotional memories are mediated by the amygdale. The temporal lobe is the seat of semantic knowledge.

The review of the neurological correlates of memory presented by Gazzaniga et al. has yielded results that again parallel those of Carter et al. In the short term memory area the lateral frontal lobes and the inferior parietal lobes are involved. Visual-spatial information seems to be stored in the parietal-occipital region while the left hemisphere is the storage area for linguistic information.

Crucial in long term memory storage is the medial temporal lobe during both storage and recall. The medial temporal lobe seems to be instrumental in triggering the regions of the brain that were involved during the original storage process. The patient, HM, suffered damage to the medial temporal lobe as a result of surgery.

Both the Carter and Gazzaniga groups note the vital role of the hippocampus in the dynamic aspects of memory. The hippocampus appears to be involved in the capturing, storage, and retrieval of information. A computer metaphor here may be applicable. In the computer there is a central processing unit (CPU) that carries out the executive functions of information processing. The CPU is involved in both the storage and retrieval of information. In storing information the CPU assigns an address to each piece of information stored so that it can be later identified and retrieved. The hippocampus seems to serve a similar role. Damage to the hippocampus can lead to severe memory problems.

## Consciousness

In almost every discipline there is a question the answer to which is endlessly debatable. In physics it might be "What happened before the big bang?" In a painting class it might be "What is art?" In music it might be "What is jazz?" In communication it might be "What is communication?" And if there were such a question in neuroscience it would probably be "What is consciousness?"

Over one century ago William James (2001) used the stream or river as a metaphor of consciousness. He said that "stream" and "river" are the metaphors by which consciousness is most naturally described. He further added that consciousness is the stream of thought or the stream of subjective life. James interchangeably used the phrases stream of consciousness and stream of thought. Fond of metaphors James compared the stream of consciousness to a bird in flight. It is like a bird in flight that occasionally perches; there is alteration between dynamic and the static states. James noted that every state tends to be part of a personal consciousness, that personal consciousness states are always changing, and that personal consciousness states are always selective; some objects or topics are accepted at any given time while others are rejected.

The neurosurgeon, Wilder Penfield (1966), found neurological support for the stream metaphor. Before doing surgery in an attempt to relieve the symptoms of severe epilepsy, Penfield would electrically stimulate areas of the exposed brain to determine the maximum amount of brain tissue that could be excised with the minimum resultant dysfunction. A by-product of the electrical stimulation was the discovery of various functions of certain brain regions. Stimulation of the anterior part of the temporal lobe, for instance, produced what Penfield called experiential responses wherein patients would relive some previous period in their lives. For example, a patient being so treated exclaimed that she was experiencing a familiar memory. Following this exclamation the patient then described an office scene from the past. Patient responses that were responses to the present, Penfield labeled, "interpretive responses." Again it was electrical stimulation that evoked these responses.

Patients never looked upon experiential responses as remembering an event. Instead they said that they were reliving the experience. Paralleling the stream metaphor Penfield noted that the experiential experiences always moved forward, never backward, and there were no still pictures. He further noted that as long as the electrode was in place the experience of a former day went forward and when the electrode was withdrawn the experience stopped. It is amazing that so many memories reside in the brain evoked sometimes only in response to electrical stimulation directly to the exposed brain by the neurosurgeon. A comparison to the computer would seem to be appropriate. In the computer there can be infinite loops that keep recycling the same information. Penfield's discoveries suggest that there are numerous infinite loops in the brain holding storehouses of past experiences that may have been seemingly forgotten.

In discussing consciousness Penfield noted that Heraclitus had the idea of consciousness when he stated that we never descend into the same river twice. William James tackled the problem twenty-four centuries later but still there was no advance in knowing about consciousness and the workings of the brain. Penfield believed that his work and the work of his colleagues was a beginning, but only a beginning, noting that: ". . . today it is just as difficult to give an adequate definition of the mind as it ever was. Consciousness is an awareness, a thinking, a knowing, a focusing of attention, a planning of action, an interpreting of present experience, a perceiving."

More than half a century later scholars are still discussing the problems of consciousness beginning with the question, "What is it?" Edelman (2006) proposed that consciousness is cortical activity between cortical regions and the thalamus, the cortex interacting with itself, and the cortex interacting with subcortical structures. In consciousness the brain speaks largely to itself, noted Edelman. Edelman introduced the concept of "qualia." Qualia are fine discriminations made by individuals such as the redness of red. Edelman was careful not to assign consciousness to a specific area of the brain. Lastly, Edelman believed that language allowed humans to be conscious of their consciousness. Robinson (2007) noted that consciousness is among the most vexing problems in both philosophy and science. As a "state," Robinson noted that consciousness seems resistant to translation into physical terms and measurements, though its dependence on a healthy nervous system appears to be as close to a cause-effect relationship as any in the natural sciences. Robinson reflected on how the issues surrounding consciousness raise questions in ethics and end-of-life issues. Finally, Grim (2008) looked at the "hard problem of consciousness" from the functionalist view giving both support and criticisms of that view. He then changes focus and looks at consciousness within the confines of physics and neuroscience. But Grim concludes that even these reductionist approaches yield unsatisfactory results.

## Neurological Correlates of Consciousness

While experiential and integrative responses are triggered by stimulating areas of the temporal cortex, Penfield (1975) said at the end of his career that the higher brain stem was the *sine qua non* of consciousness: "Gradually it became quite clear in neurological experience that even large removals of the cerebral cortex could be carried out without abolishing consciousness. On the other hand, injury or interference with function in the higher brain stem, even in small areas, would abolish consciousness completely."

Researchers in the twenty-first century would agree with Penfield. Merker (2007), for example, concluded his extensive review of the issue noting that brainstem mechanisms are integral to the constitution of the conscious state, and that an adequate account of neural mechanism of conscious function cannot be confined to the thalamocortical complex alone.

Carter et al. have identified a number of brain regions that they believe are involved in consciousness:

- The Brain Stem stimulates cortical activity without which there would be no conscious awareness.
- The Hippocampus is involved in the storage and retrieval of information. Without the functioning of the hippocampus, no consciousness would be possible.
- The Thalamus, often described as the great sensory weigh station, attenuates and directs sensory input.
- The Temporal Lobes store personal memories; the left temporal lobe is involved in linguistic processing.
- The Temporal-Parietal Junction stores the brain's map of self in relation to the physical world and reconciles the two.
- The Orbital Frontal Cortex is the seat of conscious emotion. If this area is dysfunctional, reactions to external stimuli are merely reflexive and devoid of emotion.
- The Dorsolateral Prefrontal Cortex is responsible for reconciling different ideas and perceptions; a function thought to be necessary for conscious experience.
- The Supplementary Motor Cortex is the brain region where deliberate actions are rehearsed.
- The Motor Cortex is believed to be crucial to the sense of self which would seem to be necessary for consciousness.
- The Primary Visual Cortex is vital if conscious vision is to be realized.

The first three items on the Carter et al. list are in agreement with both Penfield and Merker. The remaining items on the list are suggestions, according to Carter and her colleagues. Strong documentation has not yet been established.

# Intelligence and Thinking

Traditionally thinking and intelligence have been very much equated. People who think or who are intelligent are able to learn, are verbally skillful, are fast responders, are dexterous, and are quantitatively skillful, for example. Over the years intelligence has been assessed through testing. Two popular tests have been the Stanford-Binet and the Wechsler. The former is most often used in testing children (although its norms extend through the early twenties). The latter is used mostly with adults, the Wechsler Adult Intelligence Scale (WAIS), although there is a children's version, the Wechsler Intelligence Scale for Children (WISC).

The Stanford-Binet tests in four areas: verbal reasoning (word knowledge would be assessed here, for example), abstract/visual reasoning (completing a puzzle would be assessed here, for example), quantitative reasoning (numerical concepts are assessed here), and short term memory (a testee might be asked to reproduce a picture from memory in this part of the test). Two major areas tested in the WAIS are the Verbal and the Performance areas. The verbal area includes testing for information, similarities, and vocabulary. The performance area includes perceiving block designs, matrix reasoning, and picture completion. Response time assessment includes digit symbol coding and symbol searches.

## Neurological Correlates of Intelligence and Thinking

When one is thinking a number of areas in the brain are active. Carter et al. note that both hemispheres in the brain are active and that some activity occurs only in the left hemisphere of the brain. The frontal lobes and the parietal lobes in both hemispheres are active and there is some activity that occurs n the left frontal and parietal lobes. There is a large bundle of fibers called the arcuate fasiculus that carries messages between the frontal and parietal lobes. When a person is making inside-the-head numerical computations there is activity in the middle part of the left frontal lobe.

## Beyond Traditional Views of Intelligence and Thinking

In the latter part of the twentieth century there were reactions to the standard methods of testing a person's ability to think or a person's intelligence. These reactions were seen in a number of books, notably Dane Archer's *How to Expand Your S.I.Q. (Social Intelligence Quotient*) and Daniel Goleman's *Emotional Intelligence: 10$^{th}$ Anniversary Edition; Why It Can Matter More than IQ*. Archer (1980) focuses on social perception with a number of frames which ask the reader to answer a number of questions about pictured social scenes. For example, there will be a picture of two women and one child. The question is: "Which woman is the mother of the child?" Goleman (2006) focuses on social sensitivities. He deals, for example, with communication between spouses and how they may better communicate with more respect for each other's feelings.

In the lay world the kinds of intelligence that Archer and Goleman talk about are often called "street smarts." For example, Demos (2006) reported that the founder of Cirque du Soleil, Guy Laliberte, who began his career as a street performer and built an entertainment company which has performances running in one hundred cities, on four continents, credited his success in the entertainment business to hiring the right people. Laliberte credits much of his success in hiring the right people to his own days on the street. He said, "You can get killed pretty fast, so you develop the ability to read people."

As traditional views of intelligence have expanded, there has been a divergence in the meanings in the concepts of intelligence and thinking. Taking an intelligence test is a performance. Determining a life plan requires thinking. Thinking is a very complex process that has prompted a number of questions for researchers. The Russian psychologist, Vygotsky (1962), writing early in the twentieth century, asked if we think in words. He said metaphorically:

> A thought can be compared to a cloud which sheds a shower of words. Thought fails to coincide not only with words but also with the meaning of words but also with the meanings of words in which it is expressed, yet the way from thought to words leads through meaning. . . . If thought may be compared to a cloud shedding a rain of words—the motivation of thought might be compared to wind which sets the clouds in motion. A true and full understanding of another person's thought becomes possible only when we understand its real affective-volitional basis. (p. 533)

From Vygotsky we see that thought is opaque. We can see the results of thoughts; we can hear people speaking about their thoughts, but we do not have access to the actual thought process.

Another early twentieth century writer John Dewey (2009) spent a large part of his professional career thinking about thought. He defined thinking as "that operation in which present facts suggest other facts (or truths) in such a way as to induce belief in the latter upon the ground or warrant of the former." Dewey noted that in all fruitful thinking there is a rhythm of the conscious and the unconscious and that there is always an unconscious aspect to thinking.

Carter and colleagues (2009) note that the creative moment or the "eureka" moment seems to occur when the brain is in the alpha brain wave state (see the beginning of this chapter). Then, when the moment does occur, the brain shifts into the gamma state. This reinforces again the notion that the brain is never doing nothing. Nothing never happens.

# Decision Making

Thinking often precedes the making of a decision. In some cases there is virtually no thinking before a decision is made. You wake up to find a snake in your kitchen, you respond with virtually no thought. There is no investigation to determine if the snake is harmful or poisonous. Your amygdala kicks in and your response is purely emotional, usually fright, flight, or maybe for the brave of heart, fight. We tend to think that other decisions should be based on logic with emotion rendered out of the process. However, the neurological evidence seems to argue against this. The classical case cited is that published by Antonio Damasio (1994). The case was that of Elliot who had a brain tumor which damaged his prefrontal cortex, both left and right sides. Elliot lost his feelings and emotions; he seemed to have no internal ability to feel emotions. Most surprising, however, was Elliot's inability to make a decision. This ability was most noticeable in social situations. When Elliot did make a decision it was a poor decision.

Damasio said that Elliot seemed not to learn from his mistakes and that he seemed like the repeat offender who repents but thereafter commits another offense. Damasio reports too that in his many hours of conversation with Elliot he never saw a tinge of emotion, never saw sadness, impatience or frustration. Damasio stated that excess emotion may lead to irrational behavior, the reduction in or absence of emotion may lead to equally irrational behaviors. Damasio like others has suggested that criminal behavior, particularly psychopathic behavior, may be attributable to an inability to feel emotion or to see the consequences of one's actions.

Decision making is often divided into two categories: economic decision making and moral decision making. Economic decision making deals with rewards and costs. The immediate conflict is between immediate rewards versus reward postponement. Should you eat that second piece of chocolate cake (an immediate reward), or should you not, in order to achieve or maintain a certain health status? Researchers have devised tests to examine how people deal with immediate and delayed rewards. Damasio helped to create and certainly popularized what has come to be known as the Iowa Gambling Task. In doing the task participants are presented four virtual decks of cards on a computer screen. They are told to draw cards from the decks, that with each draw they will always win money but that sometimes they will lose money as well. Some decks are bad decks (continually drawing from these decks results in losses exceeding gains), and some decks are good decks (gains exceeding losses). Healthy participants usually learn within forty to fifty draws which are the good or bad decks, but participants with damage to the prefrontal cortex (specifically, the orbitofrontal cortex) continue to draw from the bad decks and, therefore, continue to lose.

Healthy (non-brain-damaged) participants can sometimes lose because they think too much. For example, if participants are told that on a screen the color red will flash 75 percent of the time and the color green will flash 25 percent of the time, and if the participants are asked to guess which color will appear next if the colors are presented at random, many normal participants will ignore the odds and actually try to guess which color will flash next. Gazzaniga (2008) attributes this behavior to the left hemisphere which tries to create elaborate hypotheses. The optimal strategy, of course, is to guess red every time. This is what the right hemisphere tends to do, and as a matter of fact, this is what many mammals will do in laboratory experiments.

Of course product developers and advertisers are interested in the decision making processes of consumers. Can fMRIs of a test group of consumers provide some answers? Will they buy the wine, for example? While test subjects are having their brains scanned marketers can vary the price, the appearance of the bottle, the name of the vineyard, and so on to see what neural reactions are displayed by the subjects. If the test subjects show a spike of activity in the nucleus accumbens, a brain center that anticipates rewards, and part of the frontal cortex, which is involved in balancing gains and rewards, it usually means that the test subject is likely to buy the wine. And, if the test sample of consumers really represents a much larger group of consumers, a marketing coup has occurred.

Moral decisions can be very difficult. Sapolsky (2005, Lecture 12) describes one classic problem (developed by Greene et al.) to demonstrate this; it is the runaway trolley problem. Imagine that you are on a bridge observing a runaway trolley racing out of control that is about to kill five workers on the tracks below. The workers are unaware of their imminent demise. You can pull a lever which will divert the trolley to another track which will save five lives but will kill one person on the side track. Or, you can push a person from the bridge onto the track which will kill that person but save the five lives otherwise in jeopardy. Which action would you take? In each case you will save five lives and kill one person. The outcomes are the same. Results show that three times as many people will pull the lever than will push the man. fMRIs of subjects participating in this study show that pulling the lever activates parts of the prefrontal cortex which is associated with decision making occurring as a result of logic and utilitarianism. Thinking about pushing a man off the bridge seems much more personal and emotional, and as a result, activates the limbic system in the brain. This seems to be a much more troubling choice for the participants.

## Summary and Conclusions

In this chapter a great deal of time was spent on intrapersonal communication. Why? Because intrapersonal communication is the most important form of communication. It underlies communication in all other contexts. Recall that in chapter 4 Harper's theory of communication was outlined. In that five category theory, only one category, operationalism, dealt with delivery (speaking, writing signing, and so on). The remaining categories, categorization, conceptualization, symbolization, and organization involve intrapersonal communication. A message delivered, therefore, is just the tip of the iceberg. Communication practitioners and communication scholars have been aware of this since antiquity. Neuroscientists are just coming on board helped immeasurably of course by new scanning techniques, but helped immeasurably too by the recognition that nothing never happens in the brain. More dramatically, researchers have begun to study the default mode network or have engaged in non-event studies.

The task is a daunting one. As Vygotsky noted, we can see the manifestations of thought but we cannot see thought. A brain scan might suggest that a person is thinking but the outcome of that activity is unknown. What is happening when the "eureka" moment occurs when someone is sleeping or when someone is in the alpha brain wave state? It is difficult to tell. What is known is that a gospel song writer will probably not have an "eureka" moment that will lead to a Nobel Prize in chemistry nor is it likely that a chemist will have a "eureka" moment that will lead to the writing of a new gospel song.

The issue is one of creativity and education. For the scientist the most creative moment probably occurs at the point when the hypothesis is formulated. There may be some creativity in determining how to test the hypothesis but after that things are pretty much *pro forma.* For those in the humanities, the final expression of a work of art is undergirded by a litany of private thoughts streaming through the mind of the artist.

Can creativity be taught whether it is in the sciences or in the humanities? This is almost like asking the existentialist how to experience the moment of truth or the peak experience. The most you can do, say the existentialists, is to be open to the experience occurring if it does occur. Knowledge of a discipline and a passion to answer questions posed by that discipline seem to be key ingredients.

# *Chapter 6*
# Interpersonal Communication

## Prologue

Interpersonal communication begins when life begins and possibly before. It is those early lifetime experiences that sear the mind like a branding iron on the body. Sample the content of various types of talk therapy and you will find that early interpersonal dynamics, particularly family themes, dominate the dialogues. In this chapter on interpersonal communication, we will begin at the beginning and explore those important early years from the very outset.

## Theories of Interpersonal Communication

Perhaps because of his upper-middle-class upbringing, or because of his experience with children separated from their families in post-World-War II London, or because of his belief that throughout history, genetic selection favored people who became attached to others, and as a result, survived more often than those who were not so attached, John Bowlby (1969) became very interested in mother-baby attachment. His attachment theory put forth the idea that attachment behaviors were essentially behaviors that developed to help the baby survive. If the baby and its mother were attached, then the baby had a greater chance of surviving. Babies who failed to attach to their mothers often suffered severe psychological problems that lasted into adulthood. The mother to whom the child attached served as a secure base for the baby. As the baby aged and began to explore its environment and stray with greater frequency from its secure base, the baby still knew at some level that it could return to its secure base. Bowlby believed that the secure base was most often the biological mother but that healthy attachments to other loving caregivers were also possible. Attachment behaviors may have originally evolved to protect the baby from predators but the consequences of attachment failures today are primarily psychological. While Bowlby published many manuscripts describing his theory, one major source was *Attachment: Attachment and Loss* first published in 1969 and updated and re-published after Bowlby's death in 1999.

## Bonding Theory

Klaus and Kennell (1972) published a landmark study proposing a new theory on mother-infant attachment called "bonding theory." The theory held that during the first few hours of life there is a sensitive period when close physical contact with the mother is necessary if the baby is to develop normally. The study had a dramatic influence on birthing procedures in the United States. In hospitals, deliveries of newborns occurred in environments resembling surgical theaters prior to the Klaus and Kennell article. Following the article's publication, the birthing environment became more friendly. Family members were invited in to witness the birthing process, husbands often served as "cheerleaders," and most importantly, after delivery the baby was presented immediately to its mother so the mother-infant bond could be established.

Bonding theory is not without its critics. For example, Lamb (1982) criticized that bonding theory has had little empirical support. He noted that there were too few subjects in the sample from which bonding theory was derived. Lamb also noted that the small sample of mothers used in the study were not representative of the vast majority of mothers. Lamb did not deny the possible validity of the theory, but his claim was that validity had not been established. Eyer's (1992) critique is a bit more caustic than Lamb's; she claims that bonding theory is a conspiracy, fabricated by physicians as a conspiracy against nurse-midwives who were beginning to take business away from obstetricians. Nurse midwives often delivered babies in the new mother's home, giving friendly and supportive care in a familiar environment. This was in contrast to the cold steel delivery room of the hospital. According to Eyer, bonding theory gave obstetricians a scientific rationale to change their birthing procedures, conducting them in a less harsh environment.

It is probably obvious that attachment theory and bonding theory are very similar. Both focus on the mother-infant bond at the beginning of life and early in life. If there is a difference, it is that bonding theory places greater emphasis on the first few moments of life. The bonding theory disciple observes the mother and child from the moment of birth. Bypassing any bathing, eye cleaning, or administering medicines of any kind, the newborn baby is placed on the mother's abdomen or sternum. The baby may lie there for approximately thirty minutes looking at the mother who is looking at the baby and holding the baby, often crying and smiling at the same time. The baby then begins to crawl using its legs for propulsion to the mother's breast. If the mother's breasts have been washed, this migration process is usually interrupted. The "bonding" obstetrician will usually not have the breasts washed, and if they have been, will coat them with amniotic fluid. This will trigger the baby's migration to the breast. All of these mother-infant behaviors are necessary, according to the bonding theorist if the baby is to develop normally.

Attachment and bonding phenomena are believed to have an impact throughout life. For example, Karen (1998), building upon the work of Bowlby, has noted that if a child has had a traumatizing past because of the failure of the infant-parent bond to form, a vicious cycle may repeat itself across the person's life. The person in adulthood may repeat the mistakes or failures of his or her own parents, particularly the mother. The person's children will be treated the same way the person was treated in such a way that the dysfunctional perceptions and behaviors are passed on from generation to generation.

Levine and Heller (2010) also believe that early life experience can shape interpersonal relationships later in life, giving generous credit to Bowlby and his colleagues. The authors are particularly concerned with romantic relationships. They note that researchers have long observed that children who have experienced certain attachment styles will have characteristic interpersonal styles in adulthood. More specifically, these interpersonal styles can be manifest in romantic situations. Levine and Heller have isolated three different attachment styles: anxious, avoidant, and secure. Anxious people are anxious when their partners are away, fearing that their partners may become interested in somebody else and will begin to love that somebody else and not them. Avoidant people believe that their partners want to be more intimate than they are willing to be. Intimacy seems to be equated with dependence. Finally, secure people have little difficulty expressing their needs and wants. Secure people can be comfortable when their partners are away or near. They know their partners will be back; they don't feel that their partners will abandon them.

## Neurological Underpinnings of Attachment and Bonding Theories

Chugani (2004) discussed the neurological consequences of severe cases of maternal support and deprivation in Romania in the 1980s. During this time in Romania there were serious social, political, and economic conditions. Approximately sixty-five thousand infants were placed in orphanages. Eighty-five percent of these infants were younger than one month. For infants younger than three years old there was one caregiver per every ten infants. For children older than three years old there was one caregiver for every twenty children. Conditions were deplorable. On average the infants and children spent twenty hours per day in their cribs. Many of these children were adopted by families in North America and the UK. After adoption the children experienced considerable recovery from their physical, cognitive, and social deficits. Mild deficits in these areas remained, however.

The adoptees became subjects in a number of developmental studies. Chugani compared a cohort of the adoptees with a control group; a group that had not experienced maternal deprivation. Using PET scans Chugani found that the adoptees showed decreased glucose metabolism in the orbital frontal gyrus, in the infralimbic cortex, in the amygdala, the hippocampus, part of the temporal lobe, the brain stem, and the anterior cingulated gyrus. The finding of decreased glucose metabolism suggested that the brain's response to stress in the adopted Romanian orphans had been compromised. There were variations in the adoptee group which Chugani attributed primarily to genetic factors. Chugani suggested that the extreme deprivation of the adopted orphans probably extended into adulthood. For example, even a child adopted into a loving home may have the fear that "this won't last." Or the child might continue to "test" the adoptive parents with increasingly outrageous behavior so that eventually the parents will behave in a way proving to the child that the love and happiness was a fleeting thing.

Cozolino (2006) noted that the brain is a social organ that is built through experience. He is a believer in attachment theory and how attachment has survival value for not only the baby but for the adults as well. Corzolina presents a number of case studies with adults from his psychotherapeutic practice showing how present day problems can be traced to attachment problems in childhood. But where does it all begin? Immediately after birth the baby looks at the mother's face. This leads to an increase in the baby's brain of endogenous opiates that are very instrumental in social interaction. In addition, the presence of the mother stimulates the production of corticopontine-releasing factor in the infant's hypothalamus thereby stimulating the infant's sympathetic nervous system. Also triggered is the production of endorphins and dopamine, two chemicals produced in the brain that soothe and reward. Endorphin and dopamine levels in mother and infant rise, fall, and rise again as they grow closer, separate, and grow closer again.

Cozolino believes that light touch and comfortable warmth lead to increases in oxytocin and endorphin production in both mother and infant. These are the brain-produced chemicals that enhance social bonds. Touch can also bring about mild sedation and pain relief (Think of what you do reflexively when you bump your elbow, for example. You rub it.). In addition touch can also lead to decreases in heart rate and blood pressure in both mother and infant. If a mother suffers from brain damage, particularly to the anterior cingulated gyrus, her nurturing skills will predictably be dysfunctional and the infant may indeed fail to bond.

Cozolino holds that we all have an internalized mother which consists of a network of visceral, somatic, and emotional memories of our earliest interactions with our mothers. If we fail to develop that internalized mother, there will be lifelong consequences. Adult relationships will produce stress. We will not achieve the serenity that is necessary to face the slings and arrows of everyday life. The internalized mother is developed as a result of soothing touches, being held softly and securely, feeling comforting warmth, and having repeated experiences of emotional changes from states of distress to states of calm.

## Mirror Neurons

Marco Iacoboni's (2008) rather lengthy quote suggests that every day interpersonal interactions are strategic. There is always a sizing up of the other or as Iacoboni describes it, "We read the world." Even with spouses, children, other family members, and close friends on any encounter, we tend first to read and then decide on the best strategy before we react.

> When we get right down to it, what do we human beings do all day long? We read the world, especially the people we encounter. My face in the mirror first thing in the morning doesn't look too good, but the face beside me in the mirror tells me that my lovely wife is off to a good start. One glance at my eleven year old daughter at the breakfast table tells me to tread carefully and sip my espresso in silence. When a colleague reaches for a wrench in the laboratory, I know he's going to work on the magnetic stimulation machine, and he's not going to throw his tool against the wall in anger. When another colleague walks in with a grin or smirk on her face—the line can be fine indeed, the product of tiny differences in the way we set our face muscles—I automatically and almost instantaneously can discern which it is. We all make dozens—hundreds—of such distinctions every day. It is, quite literally, what we do.
>
> Nor do we give any of this a second thought. It all seems so ordinary. However it is actually extraordinary—and extraordinary that it feels ordinary. For centuries, philosophers scratched their heads over humans' ability to understand one another. Their befuddlement was reasonable: they had essentially no science to work with. For the past 150 years or so, psychologists, cognitive scientists, and neuroscientists have had some science to work with—and in the past fifty years, a lot of science—and for a long time they continued to scratch their heads. No one could begin to explain how it is that we know what others are doing, thinking, and feeling.
>
> Now we can achieve our very subtle understanding of other people thanks to certain collections of special cells in the brain called mirror neurons. (p. 3-4)

Iacoboni attributes this behavior to mirror neurons and tells the story of their discovery. The story began almost two decades ago in the neuroscience laboratory of Giacomo Rizzolatti in Parma, Italy. Macaques with electrodes implanted in single neurons were being studied to identify the neurons' roles in various motor behaviors. While one macaque with an implanted electrode was resting between experimental trials another non-implanted macaque was cracking a nut (here various versions of the story have been given, eating ice cream or eating a banana, for example). Surprising was the serendipitous finding that the implanted observing macaque exhibited neuronal firing in its premotor cortex. In short the observing macaque's brain was performing as though it were cracking the nut. In short, it was neurologically mirroring the behavior of the macaque it was observing. Thus, began massive investigations of mirror neurons in both non-humans and humans.

In humans mirroring is of critical importance in interpersonal communication. It is the mechanism for social learning, learning that begins in the first weeks of life. Ultimately, according to Iacoboni, mirror neurons are instrumental for the transmission of culture. Iacoboni is clear to point out the difference between "mirroring" and "mimicking." Imagine a teacher-student scenario. In mirroring if the teacher raised her right arm and the student raised his left arm mirroring has occurred. If, however, the teacher raised her right arm and the student raised his right arm mimicking has occurred.

Mirror neurons are responsive to different stimulus modalities. They will fire, for example, when someone kicks a soccer ball, when someone sees a soccer ball being kicked, or when someone hears the word, "kick." They seem to respond best to the intent of an action. For instance, if an individual makes a grasping action with no discernible target for the grasp the mirror neuron response for the grasping action is less than if the observer sees the target of the grasp, say a coffee cup.

Iacoboni suggests that there is a link between mirror neurons and language. This suggestion first came to him when mirror neuron responses were observed in humans in response to action words, e.g., kick, lift, jerk, drop, hit. Gestures seem to be tied to language development beginning most likely in the first week of life. Iacoboni discusses two types of gestures: beat gestures and iconic gestures. The former assist the speaker in maintaining speech rhythm; the latter are useful to the listener by facilitating the understanding of the meaning of a message. Further evidence for the relationship between mirror neurons and language has emerged from TMS (see chapter 2) studies wherein transient lesions have been induced in Broca's area. The results of these inductions show that both speech and imitation functions are disrupted. Gestures lead then speech follows, suggesting further that mirror neurons are critical for speech and language development. The interdependence of speech and gesture dashes some cold water on the often espoused dichotomy between verbal and nonverbal communication.

Iacoboni has compared speech giving and conversation. Logically it would seem that giving a speech would be easier. A speaker can plan a speech ahead of time by producing an outline or verbatim text. The speaker can even rehearse the speech before a mirror or video camera producing a record to review. But still, unrehearsed conversation is easier. The reason, not surprisingly, has to do with mirror neurons. Speakers mirror their conversational partners' movements and syntactic structures. They will often settle on a common vocabulary and negotiate meanings. Words and actions in a conversation tend to be part of a coordinated activity. There seems to be a common goal. The dialogue partners seem to be in a dance where no one leads and no one follows.

Mirror neurons and empathy are strongly linked according to Iacoboni. When someone approaches us with a facial expression of sadness, for example, our motor neurons that are involved in the expression of sadness fire and we feel sadness and our face expresses sadness as well. Ours is an empathic response. We make similar responses to other facial expressions we observe in others. It is not surprising that people who supervise service workers stress the importance of smiling at the customer. A broad smile from the *maitre de* at a restaurant, for instance, can help facilitate a feeling of happiness in the entering diners. Of course, not all facial expressions create empathic feelings. An angry face can provoke a freeze, fight, or flight response, for example, by triggering a reaction in the amygdala. We may respond to an angry face with an angry face but we may respond also with fear. Evidence for the assertions about mirror neurons and empathy has been found in experiments where a person in a scanner is informed when a significant other, outside of a scanner, is going to be shocked. Expressions of fear and pain are seen on the person inside the scanner and mirror neurons provoked by fear and pain fire in that same person even in the absence of shock.

A not unexpected question often asked of mirror neuron researchers is, if the firing of motor neurons in the brains of others, triggers the firing of the same motor neurons in us, why do we not exhibit overt motor behaviors similar to the behaviors we observed? The answer seems to be that the strength of the observer's motor mirror neurons is weaker than the firing rate of the actor's motor neurons. A second question that arises is methodological. Can we reach the same conclusions on human studies that we can reach with macaque studies where the animals have electrodes implanted in single cells in their brains? The answer is "no." Implants cannot be placed in the brains of human subjects. It requires major surgery. Only in patients with severe brain pathologies where surgery is already indicated, epilepsy, for example, can this be attempted. Therefore the interpretation of the results of human studies requires inference, inference based on the behavior of clusters of neurons which Iacoboni calls ensembles. Some attempts have been made with macaques to simultaneously record from electrodes implanted in single cells and from fMRI scans. This is difficult since the magnetic power in the scanner burns the electrodes. Some success in solving this problem has been reported.

## Interactional Synchrony

Nearly a half century ago before there were fMRI and other modern brain scanning techniques, William Condon (Condon and Ogston 1966) discovered that whenever two people are communicating they often exhibit synchronous behaviors. The body of one can move in synch with the other's speech, or the bodies of the two communicators can move in synchrony. Condon called this interactional synchrony. Unlike the mirroring behavior that Iacoboni talked about, in interactional synchrony the parts of the body that get in synch can be mirrors or can be mismatched. For example, in one of his many films Condon showed a psychiatric interview in which a male psychiatric resident is interviewing an attractive young female patient. The resident was seated behind a desk; the patient was seated in a chair beside the desk. Each time the patient would cross her legs the resident would lean forward. Soon the leg crossings and the forward leans began to occur simultaneously. The two were in interactional synchrony. The metaphor of the dance seemed appropriate to describe this scene. Condon believed that synchrony between two individuals is a sign that rapport has been established between them. Synchrony is difficult if not impossible when a conversational participant suffers from a pathology, autism, schizophrenia, or speech and language dysfunction from a stroke, for example. Condon believed that if synchrony between individuals could not be established, then perhaps one of the individuals suffered from some pathology.

Condon further believed that interactional synchrony demonstrates a kind of behavioral identification. Two people identify with one another when their bodies move in synchrony and their intonation and vocal qualities match. Virtually every student of public speaking has learned about identification. Briefly, at the beginning of a speech a speaker must somehow identify with the audience; compliment the audience on the sports' teams the audience supports, or tell something very laudatory about the audience's role, for example. A speaker addressing a group of fire fighters, for instance, might say something about the importance of first responders in any community. For a speaker to persuade an audience the audience members must somehow believe that the speaker is one of them. The only exception to this seems to be the charismatic speaker whose mere presence overwhelms the audience.

Davis (1973), who visited Condon and spent an extended period of time in his laboratory, said that one of the most remarkable of his many, many films was the one in which two participants were wired up to an electroencephalograph (EEG) to record their brainwaves while they talked. Two cameras were used; one focused on the two participants; the other focused on the EEG tracings. Noted in this experiment was the synchrony that was established between the two participants' EEGs. Condon drew no hard conclusions from this experiment but believed that eventually the neural basis of interactional synchrony would be discovered.

## Comparing Mirroring and Interactional Synchrony

As mentioned above, Condon worked prior to the development of modern scanning techniques, particularly fMRI. Therefore comparisons must be made primarily on the basis of behaviors. Both Iacoboni and Condon saw imitative behaviors among participants in interpersonal situations. For Iacoboni the imitation was manifest in mirroring; for Condon it was manifest in interactional synchrony. For Condon, leg movements of one person move synchronously with another person's head movement, but according to Iacoboni this is not mirroring. Both researchers would recommend the use of imitation to establish either mirroring or interactional synchrony. Indeed, some therapies direct the therapist to establish synchrony with the patient or client. This is done through direct imitation. This strategy is often used with autistic children who characteristically have problems with relationships. Future research may indeed show that mirroring and interactional synchrony have common neural underpinnings.

## The Interpersonal Encounter

Person A approaches Person B. What is happening in the brain? The superior temporal sulcus in the brain analyzes the other looking for nonverbal signs that compare the self to the other. The amygdala assesses the emotionality of the situation determining whether or not an immediate flight, fight, or freeze stance should be taken. If a shortcut immediate reaction need not be taken, then the orbital frontal cortex analyzes the situation. To explain this encounter further we might consider a non-interpersonal situation. You interrupt a night's sleep at 4:00 a.m. to go to the bathroom. There in the bathroom is a live snake slithering along. For most people the amygdale immediately kicks in and there is a scream and flight reaction. Other hearty souls with snake experience might grab something and kill the snake: the fight reaction. In these cases the "snake" message is never analyzed in the orbital frontal cortex. Only in a minority of cases would the perceiver begin to analyze the situation: is it a poisonous snake, for instance?

Now when two people meet there is an instant nonverbal exchange. Judgments are made on the basis of facial expressions and body movements and postures. The face generally identifies the emotion and the body communicates the intensity of the emotion. As a case in point, if Person A encounters an angry Person B then Person's B's face may reveal the presence of anger, a squared-mouth barring of the teeth, a pulsating bulging of the neck muscles and neck blood vessels, eyebrows lowered, hard staring eyes, and nostrils flared, for example. In anger the arms are contracted since anger precedes aggression, and in order to be aggressive, the body must be contracted so it can strike out. The position of the body, particularly the arms, communicates the intensity of Person B's anger

In most everyday encounters strangers are passing other strangers. Still there are message exchanges. Goffman (1963) noted that when two strangers are approaching one another on the street they will often establish mutual eye contact at about twelve to fifteen feet then break that eye contact thereafter. This he called "civil inattention." The purpose of civil attention behaviors is for two people to acknowledge each other's existence and then to ignore each other as a sign of respect. The message is: "I respect you by ignoring you."

But it is not just with strangers that we engage in pre-verbal nonverbal exchanges. Recall the statement Iacoboni made above about his eleven year-old daughter: "One glance at my eleven-year-old daughter at the breakfast table tells me to tread carefully and sip my espresso in silence." Even between close friends and intimates there are nonverbal exchanges and consequential decisions made each time we encounter them. The consequential decisions tell us how to proceed in the encounter.

Carter (2009) outlines the neurological stages that occur if an interpersonal encounter proceeds to a conversation. A summary of the steps follows.

## *The Speaker*

- Before speaking the speaker attaches links between words and ideas and memories. The linking of words to ideas occurs in the temporal lobe.
- After linking memories and ideas to words the words are matched to sounds. This occurs in Wernicke's area on the superior portion of the temporal lobe on the left side of the brain.
- Next Broca's area of the brain kicks in. Broca's area lies anterior and superior to the temporal lobe on the left side of the brain. Here the word sounds are matched to the muscles that articulate speech.
- The motor cortex located in the posterior frontal lobe of the left side of the brain directs the articulators (mouth, lips, tongue, and velum for example) to produce the word sounds.
- Finally, the cerebellum, particularly the right cerebellum which connects to the left side of the brain, takes responsibility for coordinating the fine muscle reactions needed to produce speech. (p. 148-149)

All of the above occurs in less than a second, while the whole process listed below occurs in about one second.

## *The Listener*

- When the speaker's words are spoken the listener's auditory cortex, located superior to the Sylvian fissure, is activated. Here the speaker's message is given a first stage analysis from which information is distributed to other areas of the brain.
- The amygdala does a quick "read" of the emotional tone of the speaker's message and, if threatening, triggers an appropriate and immediate response.
- The speech is decoded beginning in Wernicke's area in the left hemisphere.
- The meaning of the speaker's message is decoded in the bilateral sections of the inferior frontal cortex.
- The flow of meaning in the speaker's speech is accomplished by relating the speaker's message to the listener's memories and knowledge.

During conversations communicators express many gestures. Iacoboni had two gesture categories as noted above. They were beat gestures and iconic gestures, the former being the "drumbeat" driving speech and the latter designed to convey meaning to the listener. Condon noted that one's body moves in synchrony with one's speech. This would be similar to, if not equivalent to, Iacoboni's beat gestures.

Iacoboni noted that a monologue should be much easier to give than to engage in dialogue or conversation. First, one can prepare and rehearse a monologue. Preparation can be only marginally helpful in a conversation. Second, in reciting a monologue, the speaker has control of the timing. A speaker in conversation does not control the timing. Third, monologue speakers usually speak in complete sentences while conversational segments are often mere fragments. Finally, in conversation the participants are always switching roles from speaker to listener. This switching skill is not required in monologue. So why is monologue more difficult than dialogue? The answer, according to Iacoboni, is rooted in mirror neurons and imitation. Iacoboni has noted that conversational partners will imitate each other's expressions, syntactic structures, and nonverbal behaviors. In addition, conversational partners will negotiate their rules for turn taking and will negotiate the meanings for the words they use. The conversational partners become one. They do a dialogical dance.

Ekman and Friesen (1969) have categorized iconic gestures, gestures designed to give information to listeners.

- *Emblems* are gestures that have a direct verbal translation. For instance a maitre de who signals a waiter holding up four fingers is telling the waiter that he needs a table for four.
- *Illustrators* are gestures that are used to augment what a speaker is saying.
- *Batons* are movements that accent a particular word or phrase. A school boy tells his father, "I really hit him" and simultaneously punches his hand, for example.
- *Deictic movements* point to an agent, place or event. People giving directions often point as they say "go up this street for four blocks."
- *Rhythmic movements* are movements that suggest the rhythm of an event. Watch the feet of almost any musician and you will know the rhythm of the music being applies.
- *Pictographs* are gestures that draw pictures in the air. A common example is the hand and arm gestures that accompany a speaker's describing a spiral staircase.
- *Affect displays* are emotional displays that often accompany facial expressions. The winner of the lottery or of an athletic competition often has mouth open and arms raised in the air signaling victory
- *Regulators* are gestures that serve to regulate interactions. Speakers, for example, will often signal that they are yielding the floor by establishing eye contact with their listeners as they finish their utterances.
- *Adaptors* reveal something about a person's personality. Self-adaptors, for instance, can be hand to body gestures. Speakers whose hands persistently go to their faces may be signaling that their utterances lack credibility.
- *Object regulators* are gestures where communicators' are touching objects as they communicate. Worry beads are, in fact, designed to be object regulators.

## Summary

Every interpersonal encounter, no matter brief or benign, triggers a neural reaction. Some reactions are immediate and provoke immediate fight, flight, or freeze response. Others involve some decision making. Even with close friends, spouses, and other family members, decisions must be made on how to respond. Often these decisions reflect the facial expressions of the other person. When an interpersonal encounter is affable the communicants will frequently begin to become harmonious in both their verbal and nonverbal behaviors. It is though the two have become one. There is suggestive evidence that the brains of the participants my respond harmoniously as well. Further research will hopefully shed more light on these phenomena, and may help us understand better dysfunctional interpersonal relationships.

# *Chapter 7*
# Communication in Wider Contexts

## Approaches Used to Examine Brain Activity

In the summer of 2008, Jeffrey Goldberg (2008) underwent an fMRI brain scan in the presence of the famous neuroscientist Marco Iacoboni (2008). Iacoboni presented Goldberg with a series of photographs while his brain was being scanned. Then Iacoboni interpreted Goldberg's neural response to each of the photographs. Following is a list of the presented photographs and Iacoboni's interpretations of Goldberg's neural responses to the photographs.

- Barack Obama: Goldberg's oribital frontal cortex lit up indicating that Obama was a source of positive emotion for Goldberg.
- John McCain: Goldberg's frontal mirror neuron lit up, indicating empathy.
- Hillary Clinton: There was activity in Goldberg's dorsal lateral prefrontal cortex, indicating that Goldberg had a conflicted reaction to Clinton.
- Mahmoud Ahmadinejad: Goldberg's ventral striatum lit up; this area specializes in processing reward. This was somewhat of a surprise given Goldberg's pro-Israel stand.
- Jimmy Carter: Goldberg showed bilateral activity in the amygdala, indicating fear and dislike.
- Bruce Springsteen: Goldberg showed activity in the ventral premotor cortex and the inferior frontal gyrus; these areas help process the emotions and actions of other people. Overall Goldberg's neural response communicated his affection for the singer.
- Edie Falco: Goldberg's medial orbital frontal cortex lit up, indicating reward, but at the same time, so did the anterior cingulated cortex indicating conflict. In sum, Edie Falco's image seems to have triggered excitement and guilt at the same time in Goldberg.
- David Bradley: Bradley is Goldberg's boss. Goldberg showed activity in the frontoparietal network in the right hemisphere, possibly indicating that

Goldberg was seeking approval of himself when Bradley's image appeared.

- Goldberg's Wife: This was a surprise for Goldberg who showed activity in the dorsolateral prefrontal cortex, meaning that Goldberg was trying to protect the privacy of his relationship with his wife. There was also activity in the medial orbital frontal cortex which is a region associated with positive emotions. Iacoboni could not explain another response of Goldberg's responses to his wife's photograph: activity in the auditory cortex as though Goldberg were hearing a sound when in fact there was no sound.

The Goldberg example serves as an introduction to this chapter on communication in wider contexts and the brain. It typifies the methodology used in most research efforts designed to examine the brain activity of individuals as they respond to messages while their brains are being scanned. The messages represent the desire to obtain objective information (information unedited by the subject) from a subject. What is really on a subject's mind?, What really are the subject's likes and dislikes?, and is a subject telling the truth? Three approaches have been used to determine whether or not people are revealing what is really on their minds. They are polygraphy, pupillometrics, and now brain scanning.

## The Polygraph

According to Egolf and Chester (2007b) the search for the real person has a long history. For example, the principles of polygraphy or lie detection are not new. The ancients knew that when people are anxious, fearful, stressed, and guilty, mouths go dry, respiration quickens, hands sweat, hearts beat faster, and muscles tense. In fact, some ancient lie detection methods were based, for instance, on dryness of the mouth. A suspect whose tongue was burned by a hot sword placed on it was judged to be lying. A suspect who could not spit out dry rice powder or swallow dry bread was judged guilty.

Of course, the polygraph used today is much more technical and precise. Still the accuracy data on polygraph testing varies widely. Accuracy figures vary between 60 and 90 percent. So much depends upon the experience of the polygrapher, the gravity of the questions, and the sophistication of the subject. The question often asked is, "Can the polygraph be beaten?" A common technique is for the subject to put a tack in his shoe. Then when the control questions (questions for which the polygrapher knows the answers) are asked at the beginning of the session, the subject presses his heel down on the tack to cause pain. This triggers more intense responses in the physiological measures being tracked, usually blood pressure, galvanic skin response, and respiration. Then when the relevant questions are asked the subject releases pressure on the tack so that elevation in the intensity of the physiological measures caused by lying will be indistinguishable from that observed in the control question stage. Of course, an experienced polygrapher will look for such deception strategies in a subject.

Former CIA operative Aldrich Ames (now sitting in a jail) apparently had a very sophisticated method of beating the polygraph. He was polygraphed many times and each time passed the test. He was suspected of revealing the names of CIA operatives in Eastern Europe to the Russians and as a result many of those operatives were executed. Apparently the Russians had given Ames drugs that allowed him to beat the polygraph. What led to Ames' downfall reportedly was the fact that he was living far beyond his income and could not explain the source of extra income; that income it was determined came from the Russians.

Polygraph use has become more limited in recent years. The Employee Protection Act of 1988 generally prohibits polygraphy use for pre-employment interviews and during the course of employment. The concern is with the frequency of false positives. If a person is telling the truth, but the polygrapher determines that the person has lied, then a false positive has occurred. In testimony to the United States Congress that led to the passage of the 1988 Act the reported occurrences of false positives was very persuasive. For example, a store manager is suspected of stealing store merchandise, and a polygrapher supports this suspicion. The store manager is fired; his life is essentially ruined. However, the thefts continue after the manager's departure. Later the real thief is found and apprehended. However, it is too late for the store manager to regain his reputation. The manager was the victim of a false positive. The United States Congress was also influenced by the type of questions often asked of employees either in their job interviews or during their employment. For example, questions about an interviewee's sex life were deemed inappropriate in a routine job interview.

Polygraph use in the courtroom is severely limited because suspects need not testify against themselves, and because it can be argued that taking information from a person that is not voluntarily offered is a form of search and seizure. In the intelligence agencies in the United States government, polygraph interviews remain legal, however.

## Pupillometrics and Eye Tracking

Pupillometrics refers to the role of pupil size in communication. Eckerd Hess (1965, 1975) has been the leading researcher in this area. His research was prominent in the second half of the twentieth century. The pupil, the "hole" in the eye, expands (dilates) and contracts in response to the level of illumination in the environment. The pupil is surrounded by the iris, the colored part of the eye. Hess was aware of the primary function of the pupil, controlling the light that enters the eye. He was aware of the sensitivity of the pupil to changes of illumination. At the same time he believed that the pupil functioned not only at this primary level but at another level as well. Hess believed that, in general, when people see someone or something that they like their pupils will dilate. On the other hand, when people see someone or something that they do not like their pupils will contract. Hess conducted a number of experiments to support his beliefs. He believed that two people can communicate by mutual changes in pupil size and not be aware of it. He stated that he had found that pupil size serves as a signal between two individuals usually below the level of awareness.

Possible uses of Hess's findings can be found, for instance, in advertising. Suppose that a company wants to sell a new sun block. Let us imagine an advertisement in which a very attractive bikini-clad woman is walking on the beach holding a container of the new sun block and talking about its positive aspects. People look at this ad while a camera looks at their pupils and while another camera tracks their eye movements. What has often been found in studies like this is that men's pupils dilate at the sight of the woman, and their eyes track over the woman's body. They never see the product. Women, on the other hand, show pupil contraction when they look at the model and show little eye tracking. Neither gender shows even a threshold level of recall for the product being advertised. Men, it is believed, are attracted solely to the woman's body and fail to see or think about the product, while the observing women seem to be turned off by the ad because they feel that no matter how hard they try, they will never be as attractive as the woman in the ad. The ad agency then prepares a second ad where the same model is on vacation with her children and is reminding them that they need sun block before they can go out on the beach. Product recall is increased with the second ad.

A potentially more gravitas application of Hess's findings stemmed from the fact that before 1973 homosexuality was viewed as a psychopathology by the American Psychiatric Association (APA). This view was listed in the association's Diagnostic and Statistical Manual (The DSM). Since Hess found that male homosexuals gave different pupil responses to a variety of visual stimuli than did male heterosexuals, it was suggested that pupil responses could serve as an evaluative tool to diagnose homosexuality. In 1973, however, the board of the APA declared that homosexuality was not a psychopathology. The full membership of the APA ratified this decision in 1974 and homosexuality was removed from the DSM.

Hess suggested that pupil size may have an impact on many everyday encounters. For example, poker players are very protective of their eyes. They will wear duck bill caps or sunglasses during play to not reveal that they have a good hand via pupil dilation. Night clubs routinely turn down the lights. Is this to save electricity? Probably not. What the dimness does do is dilate the pupils of the patrons, many of them romantic couples. In essence they are telling each other, "I like you." Related to this is the fact that one of the oldest makeups for women was *belladonna* which dilates the pupils. Why, are babies so lovable asked Hess? Because they have large pupils, he answered. On the other hand, as people age, the pupils become smaller, and old people are often seen as having little beady eyes. This can be an unfortunate byproduct of aging.

## fMRI and Deception

Margaret Talbot (2007) wrote a comprehensive report on the possible use of fMRI to detect deception. She interviewed a number of neuroscientists as well as the CEOs of two companies that sell fMRI services for the sole purpose of identifying liars and their lies. Talbot compares and contrasts fMRI use in detecting deception with other methods, particularly polygraphy and pupillometrics. Polygraphy and pupillometric methods track the physiological or the emotional conflicts that result from lying. With fMRI there is conflict as well, but it is assumed that this conflict is cognitive, causing the liar to think a little harder to make up a story. This cognitive conflict triggered by lying seems to be revealed in at least three areas of the brain: (1) the anterior cingulate cortex which is associated with elevated attention and error monitoring, (2) the dorsolateral prefrontal cortex which is associated with behavioral control, and (3) the parietal cortex which is associated with sensory input. Troubling for the field, however, is the fact that some researchers have found that liars have heightened activity in other areas of the brain beyond the three areas listed above.

Critics of fMRI lie detection research have pointed out a number of areas of concern. First, the data has been generated in highly controlled laboratory experiments which often have little or no relevance to everyday life. For example, lying or truth telling about the face of a playing card has little relevance to lies told in interpersonal, social, and organizational relationships. Second, few peer-reviewed studies have been conducted on deception and fMRI. Little is known about individual differences within and across groups because few studies have been done. Third, asking experimental subjects to lie when a certain playing card is displayed is not a lie having equivalence to a guilty party (an employee of a bank) denying that he ever took money from the bank or any of its customers. Fourth, it is difficult sometimes to determine, what is a lie? For example, an ad for an automobile is seen it the classifieds. The ad says the price is 7,500 dollars or best offer. A potential buyer offers 6,100 dollars. Is that the highest sum the potential buyer is willing to pay? Is it a lie if the buyer knew from the outset that she was willing to pay 6,900 dollars for the car? Many people would answer, "No, this is just bargaining." Fifth, lie detection methods have problems with conflict. For instance, if a person is asked, "Do you love your spouse?" Well maybe on the very day that question is asked the questioned spouse had a little spousal spat at breakfast. Later when the question is asked there is still a bit of afterburn from the breakfast spat which clouds the answer "yes" to the "Do you love your spouse?" question. Last, many critics reluctantly agree that colored fMRI images of the brain are very persuasive and it may be that the persuasive effects of the images themselves that persuade a lay person who is being questioned as a suspect that the conclusions drawn from the images are valid.

Two companies that have formed to sell lie detection services are No Lie MRI and Cephos. Talbot reports that the No Lie MRI website claims that its technology represents the first and only direct measure of truth verification and lie detection in human history. One of No Lie's earliest successes was a man who was accused of deliberately setting fire to his deli. The judge dismissed the charges against the man, but he still wanted to sue his insurance company to receive the settlement he felt that he deserved. So he hired No Lie MRI and he passed their test and won his settlement. Cephos seems to be proceeding more cautiously than No Lie MRI. The Cephos CEO believes that the biggest customer pool for Cephos services will be people charged in civil cases, about fifteen million per year in the United States. These people often want to clear their names. Just as DNA tests have freed many prisoners, so might fMRI tests free or at least clear the names of individuals accused in civil cases.

Psychiatrist Daniel Carlat (2008) interviewed executives at both No Lie MRI and Cephos. Cephos CEO Steven Laken explained the rationale behind fMRI lie detection saying that when you lie your brain has to work a little harder. First you have to stop yourself from telling the truth, and second, you have to construct your lie. The fact that the brain works harder on lying than in truth telling can be detected on fMRI brain scans. At Cephos Carlat participated in the lie detection exercise from which Cephos derived its 90 percent accuracy figures. In the exercise participants were asked to "steal" either a watch or a ring from a drawer. Carlat stole the watch. Then into the fMRI scanner Carlat went. Carlat said "no" when asked if he stole the ring and he said "no" yet again when he was asked if he had stolen the watch. The analysis of the fMRI data "said" that Carlat had stolen the ring and the analysis was wrong since Carlat had stolen the watch. Carlat is skeptical about the use of fMRI analyses in neuromarketing. He concludes his report with this statement:

> Most neuromarketers are using these scans as a way of sprinkling glitter over their products, so that customers will be persuaded that the pictures are giving them a deeper understanding of their mind. In fact, imaging technologies are still in their infancy. And while overenthusiastic practitioners my try to leapfrog over the science, real progress, which will take decades, will be made by patient and methodical researchers, not by entrepreneurs looking to make a buck.

Similar to entering polygraph results in the courtroom is the situation with entering fMRI results in the courtroom. Saenz (2011) has reported on the failure to use fMRI lie detection data in a United States Court in Brooklyn, NY. At the time Saenz filed his report no fMRI data had been allowed in the determination of guilt or innocence. Such data have been used in the sentencing phase where guilt has already been established. In the Brooklyn case the defense wanted to support the credibility of a witness by providing corroborative fMRI evidence of the witness's credibility. The court said that juries, not machines, are responsible for detecting lies. If the fMRI evidence were admitted, the prosecution most likely would have presented an expert who would have testified that the 90 percent accuracy reported by fMRI companies was inaccurate for a variety of reasons. Just as with the polygraph, if fMRI use in the courtroom did occur, there would be similar arguments made for very limited use. First, suspects need not testify against themselves. Second, suspects are protected against search and seizure. Use will probably be restricted to the sentencing phase (see the chapter on Ethics below) and to cases where a suspect will attempt to show innocence from taking an fMRI lie detection examination. But in such cases the opposing attorneys will not only question the suspect but will challenge the credibility of the company that did the testing.

# Neuromarketing

## Coke v. Pepsi

It began as a TV commercial where blindfolded subjects were asked to taste two different soft drinks. The majority chose Pepsi over Coke. However, when the subjects were told which brand they were drinking the majority selected Coke. This experiment has been repeated many times in high school science classes and in other venues. It is an experiment that has bolstered the recommendations of many marketing consultants who have stressed the importance of branding to their clients. McClure et al. (2004) did the experiment while subjects were being scanned in an fMRI scanner. Plastic tubes were configured so that they could squirt sample tastes of soft drinks (Coke and Pepsi) into the mouths of subjects while their brains were being scanned. There were two key trials. In one, the brands of the drinks were not revealed to the subjects; in the second trial, the brands were identified. When the drink brands were not identified, the subjects preferred Pepsi. In this trial the ventrolateral prefrontal cortex, one of the brain's reward centers, lit up. But, when the brands were identified, 75 percent of the subjects showed a preference for Coke. Here the brain areas that lit up were the medial prefrontal cortex and the hippocampus. These areas have been identified with the sense of self, one's life experience, and memories. These experiments show that taste can be modified by culture and life experiences; in marketing lingo, "branding" is the term that is often used. In short, the brain communicates and persuasive campaigns can change what the brain communicates.

## De Gustibus Non Est Disputandum

Somewhat contradicting the Latin phrase, *De Gustibus Non Est Disputandum* (There is no accounting for taste), is the work of Plassmann et al. (2007). Twenty people tasted five wines delivered to them through plastic tubes while the subjects' brains were being scanned in an fMRI machine. The subjects were lied to because in fact there were only three different wines, but the wines were presented as five different wines differentiated only on price. For example, the subjects were told that one wine cost five dollars a bottle; at the same time this same wine was also presented as a 45 dollars a bottle wine. The results showed that subjects tended to judge the quality of a wine by its price. Quality correlated with price. The fMRI brain scans of the subjects showed that the more expensive wines made parts of the prefrontal cortex light up. When the subjects later were served the same wines with no price information, it was found that they preferred the cheapest of the wines. Subsequent to the main experiment the researchers performed the same study with members of a wine club and found similar results.

## Buying Decisions

Using fMRI scans Knutson et al. (2007) studied the brains of subjects as they considered buying or not buying a product. In his research he examined product presentation, price display, and decision making which included the price the subject was willing to pay. Knutson found that product preference correlated with the activation of the nucleus accumbens (NACC) which is related to reward processing. The NACC is part of the striatum which has cortical projections from the limbic and paralimbic cortex. The NACC is responsible for converting motivation into action. Activity here can predict the purchase of the product. When the price of the product was lower than the price subjects expected, there was activity in the medial prefrontal cortex. This fits with the data that show the medial prefrontal cortex tracks the difference between expected and actual reward outcomes in monetary reward tasks. Increased activity in the insula has been associated with aversive events whether they be monetary loss or looking at unpleasant or aversive pictures. Activity in this area has been associated with non-purchases.

## Neuromarketing Research with EEG Scans

Taking a somewhat different approach to neuromarketing is NeuroFocus. Penenberg (2011) interviewed the CEO of NeuroFocus, A. K. Pradeep. NeuroFocus is wholly owned by Nielsen Research. NeuroFocus uses a product called Mynd which is a portable and wireless EEG scanner. The scanner has several dozen sensors which pick up neural signals and sends those signals via Bluetooth to a computer which separates the signal from the noise. NeuroFocus's EEG device eliminates the restricted environment of the fMRI and the expense of using that device. Test consumers can wear the NeuroFocus helmet while watching TV at home. Some clients of NeuroFocus have been Pepsico's Frito-Lay division, CBS broadcasting, ESPN, Intel, California Olive Ranch, and ebay. As mentioned in an earlier chapter, the advantage of EEG is that it detects brain activity almost instantly while fMRI takes from one to five seconds. The weakness of EEG, however, is that it is less accurate on the location of the activity. Examples of findings from NeuroFocus's research are the findings that consumers like curves and have an aversion to sharp objects, they like to hear about the convenience of a secure internet payment service and want to hear nothing about the payment process, and that they like to see faces on models that are mysterious and ambiguous. Penenberg notes in his article that the NeuroFocus technique is not without criticism, criticism aimed at neuromarketing firms in general. One critic said, "These corporations share the same goal: to mine your brain so they can blow your mind with products you deeply desire."

## Neuromarketing Research and Government Regulation

The neuromarketing research has not escaped the overview of the United States Federal Government. In 2004, the United States Office for Human Research Protections (see Diagnostic Imaging Online, February 17, 2004) decided that fMRI studies used to help companies develop marketing strategies do not violate federal regulations. The decision was made in response to marketing research being done at Emory University in Atlanta. The issue was initiated by a group called Commercial Alert. The group believes that such research will lead to obesity, diabetes, tobacco addiction and alcoholism, for example. The research done at Emory was being done in conjunction with the BrightHouse Neurostrategies Institute, which itself was a partner with a commercial marketing firm. In a "fruits and vegetables" study, subjects in the Emory study showed activation of the medial prefrontal cortex distinguished strongly liked items from other lower levels of preference. Moreover, ventral areas in the brain associated with reward expectancy also showed activation to strongly preferred items. Less preferred items were associated with striatal activity. The government noted that review boards do not have to consider long-range effects of applying knowledge gained in the research as among those risks that fall within the purview of its responsibilities. The general conclusion of the review group was that the Emory research was descriptive and not prescriptive.

## Neuropolitics

Is selling a product different from selling a political candidate? The work of Drew Westen (2007) suggests that it is not. In his book, *The Political Brain,* Westen recruited fifteen committed Democrats and fifteen committed Republicans before the 2004 United States presidential election. The subjects' brains were scanned as they looked at slides showing statements made by their respective candidates before the election. Subjects saw a statement by their candidate followed by a statement by their candidate that contradicted the earlier statement. The subjects were then asked if they noted the inconsistency between the first two statements and finally were asked to rate the extent to which their candidates' statements were contradictory. Included too as a control were statements made by two men who were not political candidates. Westen posited four hypotheses for the study:

- Threatening information, even if the partisan subject did not acknowledge it as threatening, would activate neural circuits in the brain associated with negative states.
- Brain areas heavily involved in regulating emotion would be activated. In short, emotion would trump reason.
- Brain regions involved in monitoring and regulating conflict would be activated.
- Brain regions associated with reasoning would not be activated.

Westen found support for each of the four hypotheses. Subjects had no trouble seeing the contradictions in the statements of opposition candidates, but for their own candidates, the extent of the contradictions were perceived to be much less extensive. For the control subjects no differences were found between the contradiction ratings of the two partisan groups. Westen concluded that emotion is the key driving force in politics.

A thesis of Westen's is that Democrats use too much reason in persuading voters while Republicans use emotion. And, according to Westen, emotion trumps reason. In support of his view Westen provides a number of examples primarily from the United States 2004 presidential campaign. He also brings in brain regions in almost metaphorical ways. For example, the dorsolateral prefrontal cortex is active when people are making conscious choices, consciously thinking, and weighing evidence. This is the part of the brain to which Democrats target their appeals. On the other hand, Republicans target the ventromedial prefrontal cortex which is involved in emotional experience, social and emotional intelligence, moral functioning, and the linking of emotion and thought. Westen has called the dorsolateral prefrontal cortex the blüe brain and the ventromedial prefrontal cortex the red brain with Democrats appealing to the former and Republicans to the latter.

Shortly after Westen's book was published the linguist George Lakoff published *The Political Mind* (Lakoff 2008). Lakoff was impressed with the Westen treatise. Lakoff agreed with Westen that emotion trumps reason when voters make political decisions. He, moreover, agreed with Westen that emotion-based decisions are not irrational but that emotions are necessary to make a decision. Both writers agree with the views of the neurologist, Antonio Damasio, that people cannot make decisions if the emotional centers in their brains are destroyed. But the larger question for Lakoff and Westen is, why do partisan voters ignore the inconsistencies and contradictions in their own candidates and not in opposing candidates? Lakoff draws upon his linguistic expertise to address this question. For example, take the "entitlement" label. "Entitlement" suggests that some benefit like social security is deserved whether it has been earned or not. Thus people who receive entitlements can be demonized. These people are getting something that they think they deserve, but in fact they do not deserve it; they did not earn it.

But Lakoff is interested in more than just words. He is interested in narratives and frames. What narrative or frame underlies a voter's tendency to ignore the flaws of a particular candidate? Party affiliation can be a major factor. Partisan voters will vote for candidates belonging to their party regardless of their flaws. And candidates belonging to their party may fit a particular narrative. For example, one narrative is that individual responsibility is a high priority. Individuals control their own fates. An opposing narrative is collective responsibility. You are your brother's keeper. Lakoff, an acknowledged progressive, believes that health care should be free and available to everyone just as the streets and most highways are free even for businesses that use public motorways to make a profit. When you call the fire department do they first ask for your fire insurance card or number? No. Then says Lakoff why should they ask for your health insurance card when you go to a hospital? The reason we accept restricted hospital access is that someone has created a narrative in our minds about the situation.

## Summary

This chapter began with a description of the brain scan of Jeffrey Goldberg. Think back to that description and how the scans could be used in gaining information on Goldberg's political views, his views on media celebrities which might be used by media marketers, his views about his boss and his wife. If additional well chosen subjects were tested, generalizations could be made to a larger pool from which these subjects were drawn. This is the basis of neuro lie detection, neuromarketing and neuropolitics.

# *Chapter 8*
# Communication Disorders

The brain is complex, and it is also fragile. So many things can go wrong. It is amazing that so many people are normal. In this chapter selected brain disorders will be examined. In particular, brain disorders that affect communication will be discussed. In chapter 3 one methodology for studying the brain was to look at brain pathologies. There first was a division between physical pathologies and psychopathologies, the former showing definite tissue damage and the latter causing dysfunction in the absence of any clear tissue damage. The system for categorizing physical pathologies in chapter 3 will in general be used here. In short, physical brain pathologies were categorized into autoimmune, congenital, degenerative, infectious, neoplastic, traumatic, and toxic. The boundary between any two physical pathologies can be somewhat porous in that some physical pathologies seem to occupy more than one category. Information about the pathologies discussed in this chapter can be found in various textbooks on communication disorders, Mildner (2008), Carter et al. (2009), and The Diagnostic and Statistical Manual IV (updated in 2000).

## Physical Pathologies

### Autoimmune

Multiple sclerosis is an autoimmune disease that affects the brain and spinal cord (central nervous system). Specifically, the body's own immune system seems to attack itself and destroy the myelin sheath that "insulates" the axons of nerve cells; this insulation permits the speedy transmission of neural impulses to the muscles of the body. Without the insulation there can be a lack of coordination and motor control, for example. Eventually the neurons themselves die. The cause of the disease is not known although some genetic factor may be involved. The disease attacks more women than men. Symptoms can vary; the location and severity of each attack can be different. Episodes can last for days, weeks and longer. And equally variable are the periods of remission. There is no cure for this disease so management is crucial. In terms of communication there are first the nonverbal deficits caused by signals from the brain not reaching the muscles. This prevents the establishment of any synchrony between the sufferer and the sufferer's conversational partner. At advanced stages of the disease the sufferer can display moderate to severe speech problems.

## Congenital

A congenital disorder is one that exists at birth. Cerebral palsy is a common congenital disorder which can affect muscle movement and speech and language. There are four major categories of cerebral palsy. *Spastic cerebral palsy* is characterized by stiff movements due to tight muscles. *Athetoid cerebral palsy* is characterized by involuntary writhing movements especially in the facial area. *Ataxic cerebral palsy* is manifested in shaky movements in the hands and feet. *Combination cerebral palsy* is characterized by a mixture of the symptoms of the previous three types. Posited causes of cerebral palsy are varied. Included are problems that lead to premature birth, oxygen deprivation to the fetus before or during birth, infections transmitted to the fetus from the mother, and blood incompatibility between mother and fetus, for example. The brain damage in cerebral palsy is permanent. There is no cure for this condition.

Nonverbally, cerebral palsy victims are stigmatized before they even begin to try to speak because of their lack of muscle control. Sometimes even an ordinarily benign handshake becomes difficult. First, there might be arm flailing, causing the cerebral palsied individual to repeatedly miss the mark. Secondly, in some cases a handshake initiated cannot be terminated. The cerebral palsied individual's hand locks the hand of the conversational partner. When speech is initiated it is often dysarthric (slurred and labored) making it difficult to understand. Adding to the difficulties of many cerebral palsied communicators is a problem with saliva. Many of them drool when attempting speech. The rhythm of conversation is broken. Because of the speech and language problems, which can be severe, it is often difficult to assess the intelligence of cerebral palsied individuals. What confounds the issue is that many cerebral palsied individuals have not had ideal early educational experiences. For many cerebral palsied individuals communication is augmented with some of the devices discussed in the next chapter.

A second congenital pathology having communication ramifications is *Down syndrome.* Down syndrome is caused by a chromosomal abnormality;

there is an extra chromosome and there is no known reason for the presence of this extra chromosome. Some possible contributing factors may be the ages of the mother and the father, genetic factors may be involved since parents who have one Down child have a higher risk of having a second child with Down than parents with a non-Down child who have one or more additional children.

Although there are wide individual variations in severity of symptoms, Down syndrome children often show slow motor and language development. The appearance of the Down syndrome person is stigmatizing, characterized often by a small face with upward sloping eyes, a flattened back of the head, a short neck, and short stature overall. This appearance can immediately communicate a negative nonverbal perception in the mind of any potential communication partner. In short the Down syndrome communicator is at a disadvantage because of the negative nonverbal messages of appearance before speech is even attempted. When speech is attempted there can be deficiencies because of intellectual problems, and because of motor problems, the intelligibility of speech can be affected.

## Degenerative

Three degenerative diseases that often result in communication problems are Parkinson's, Huntington's, and Alzheimer's.

Parkinson's disease is caused by the degeneration of cells in the substantia nigra region of the brain. It is the cells in the substantia nigra that produce the neurotransmitter, dopamine. Dopamine, in addition to its role as the reward neurotransmitter, is also responsible for muscle movement. In Parkinson's there is a reduction in dopamine production. As the disease progresses, there are often tremors in the hands and legs even at rest. It becomes difficult to take a first step as well as subsequent steps, since voluntary movements are affected. As the disease progresses further, more movements are affected. Handwriting can become illegible and facial expressions may be affected. Ultimately speech may become unintelligible, the patient may become rigid, and the patient may become depressed and demented.

The first communication pathology displayed by the Parkinson's patient, as in many neuromuscular pathologies, is in the nonverbal realm. With tremors and disturbed voluntary movements the Parkinson's patient immediately sends a negative message to a potential conversational partner. Body movements and postures, known in the nonverbal literature as kinesics, are very instrumental in communication. Again, before anything is said, perceptions about a person are formed and often they are extremely difficult to erase. As mentioned in the advanced stages, speech of the Parkinson's patient is affected and intelligibility can be reduced. In the last chapter of this book (Future Trends) there is further discussion of this disease.

In Huntington's disease (often called Woody Guthrie's disease and named after the folk singer who suffered from this inherited condition) neurons in the brain degenerate, resulting in jerky muscular movements, facial grimaces, speech problems, and ultimately dementia. There are often problems with memory, ability to concentrate, and personality. Parents-to-be can be genetically tested to see if they carry the gene that results in Huntington's. Such tests are particularly crucial if one of the parents is suspected of being a carrier of the gene.

The Huntington's sufferer's initial communication problems are nonverbal. The disturbed movements immediately stigmatize the sufferer. As the disease progresses, speech problems can surface, and with further progression, personality problems can surface. The sufferer can appear to be antisocial and aggressive. If the demented stage is reached, verbal communication may be grammatically intact but it may at the same time be incoherent.

Alzheimer's disease discussed at various points in this book is one of the most feared illnesses of our time. Alzheimer's is considered to be a disease of old age, and, as mentioned earlier, one reason for this fear is the increased life span of individuals living today. Birthday parties for centenarians are at an all time high. Alzheimer's is believed to be caused by the build-up of amyloid plaque between the hippocampus and various memory centers in the brain. Although as reported in the last chapter in this book, the confounding factor in this attribution is that some individuals have the amyloid plaque but do not have Alzheimer's. There are many efforts aimed at predicting the onset of Alzheimer's, preventing it, and treating it.

Alzheimer's disease progresses in stages. In the beginning there are rather benign memory problems that may be indistinguishable from the memory problems of getting old. Then the memory problems become more severe. Sufferers have difficulties remembering recent events and may show confusion about time and place. There may be, for example, word finding problems which often lead to verbal circumlocutions. For instance, the sufferer who may not be able to think of the word, "store," may say, "You know the place where we get bread." In the final stages of Alzheimer's, the sufferer may experience severe confusion, psychotic symptoms, delusions, hallucinations, and dementia.

The chief communication problem presented by the Alzheimer's sufferer is to be found in the dementia phase. Because of severe memory problems the Alzheimer's sufferer will often speak in utterances not tethered in reality. For example, a woman suffering from Alzheimer's is asked, "What did you do today?" The woman answers that she had lunch with her husband. The problem is that her husband died eleven years ago. It is at this point that clinicians differ on the response to this irrational reply. Some say that the clinician should validate the woman's response by saying, "And, how was lunch?" This keeps the flow of conversation going and validates the world in which the woman is living. Other clinicians, on the other hand, would seek to bring the woman back to reality by reminding her that her husband was dead, that in fact, he died eleven years ago.

## Infectious

Encephalitis is a viral infection of the brain. The infection causes the brain to swell, compressing the brain against the skull. Encephalitis symptoms can range from mild to severe. In the former case there may be only a slight fever and a headache. In severe cases there can be a variety of symptoms including nausea, vomiting, muscle weakness, poor coordination, confusion, and seizures, for example. In severe cases there can also be memory loss and speech and language problems. The communication problems can therefore be both nonverbal (kinesic in nature) and verbal (speech and language problems). The sufferor's difficulties with speech and language may be due to the memory problems so speech can become unintelligible or totally absent. In the most severe cases encephalitis can cause permanent brain damage or even death.

## Neoplastic

Neoplastic (new growth) pathologies in the brain refer to tumors that can be benign or malignant. The tumors may have their origin in the brain itself or they can migrate to the brain from other sites in the body through metastasis. Other sites of origin could be the lungs, kidney, breast, or colon, for example. Symptoms of a brain tumor may be persistent headaches, dizziness, muscle weakness, blurred vision, and speech problems. In addition there can be marked personality changes. The proper gentleman at work and in social situations begins to tell off-color jokes or the Eagle Scout becomes a mass murderer. chapter 11 addresses some of these issues. Tumors can appear at any place in the brain, so the symptoms can vary widely. Correspondingly, communication problems can vary widely as well. A common reaction to individuals suffering from brain tumors is that they are not the people they used to be.

## Traumatic

Trauma to the brain can be endogenous (internal insults) or exogenous (external insults). A primary cause of endogenous insult to the brain is the stroke and key causes of exogenous insults are falls, automobile accidents, war injuries, violent acts of various kinds (gunshot wounds, fights, etc.), and sports injuries, for example. In a stroke, neurons are starved of oxygen and nourishment because there is a blockage of a blood vessel (ischemic stroke) or because a blood vessel has ruptured (hemorrhagic stroke). In the former situation prompt treatment with a clot buster drug can generally ameliorate the effects of non-hemorrhagic stroke. Strokes can occur in any part of the brain. As such, their effects can be vary in range and severity.

A stroke in the dominant or language hemisphere of the brain (for approximately 97 percent of the population this is the left hemisphere) often results in a language disorder called aphasia. Three common types of aphasia are receptive or Wernicke's aphasia, expressive or Brocas's aphasia, and connective aphasia. Wernicke's area of the brain is located in the back part of the temporal lobe. When a stroke damages this area of the brain the stroke victim has great difficulty decoding speech. In an interpersonal situation a response to a conversational partner may be inappropriate in nature because of a breakdown in the understanding of conversational speech and language. Broca's aphasia is characterized by great difficulty in formulating expressive utterances. A Broca's aphasic patient might be able to respond to a simple vocal command from another person but be unable to give a similar command in return. Connective aphasia is caused by damage to the connective fibers (the bundle is called the fasciculus arcuatus) that connect Wernicke's area with Broca's area. Messages are received and perceived in Wernicke's area but they cannot be transmitted to Broca's area. They are essentially not received in a way that permits the sufferer to intelligibly respond to them. Stroke patients can experience problems with reading and writing and can exhibit motor or muscular speech problems, slowed slurred speech (dysarthia) for example. In addition, a condition called apraxia can exist. Here there seems to be no problem with the motor speech mechanism but with the "computer" that sends signals to the speech mechanism. These problems too are based in the left hemisphere areas of the brain. Since the motor neurons that innervate the muscles of the body cross over to the opposite side of the body as they descend from the brain, stroke victims who present serious speech and language problems may also exhibit weakness or paralysis on the opposite or right side of the body.

Strokes can also damage the non-dominant hemisphere of the brain. The right hemisphere has often been referred to as the nonverbal hemisphere. As such, damage to this hemisphere often affects nonverbal processes. Prominent, is the reduced ability of right-brained-injured sufferers to perceive emotions. Emotions can be communicated by facial expressions; body movements and postures; loudness, prosody, and rate of word production; gestures; and proximity, for example. Recognition of faces and melodies also seem to be right brain functions. People whose brain damage is limited to the right hemisphere may be able to understand and produce intelligible speech, but in an extended interpersonal encounter something will seem to be wrong. Overall, both hemispheres are needed to be a competent communicator.

Exogenous insults to the brain can produce many of the same effects as endogenous insults. For example, a victim of a gunshot wound to the head can exhibit speech and language problems as well as weakness or paralysis in the extremities. Garnering a great deal of attention today is the closed head injury. Falls are the most common cause of closed head injuries but the most attention today is given to closed head injuries suffered in war zones and in athletic competition. In a closed head injury the skull and the membranes surrounding the brain stay intact but the brain is still injured by the blunt force of an object hitting the head or the sudden force of increased atmospheric pressure caused by an explosion. Many soldiers returning from war zones have suffered closed head injuries. Some of the early symptoms they exhibit are headaches, dizziness, nausea, and slurred speech. Later there may be dramatic changes in personality. Victims may become depressed, anxious, or unable to concentrate, for example. Athletes who have suffered closed head injuries have reported similar symptoms. Now the long-term effects of repeated concussions (a concussion is a form of closed injury) are being investigated. Sports fans love a good hit, particularly in football and hockey. Now it appears that those good hits come at a price. The effects on a soldier of an improvised explosive device (IED) may be more immediate, but the same effects may appear eventually in the athlete as a result of an additive effect. With closed head injury, communication problems may first be noticed in the speech of the injured person. As time goes by, speech and language problems can become more marked because of memory problems and personality changes.

## Toxic Pathologies

There are a number of substances that poison the brain. These substances are often called neurotoxins. Some neurotoxins are found in the environment and have been used in a number of materials before their toxic effects were known. Lead, for example, was once used in making dinnerware for the upper class citizens in ancient Rome and in makeup for upper class women in England past. In the United States, in more modern times when well-dressed men wore hats, hats were cleaned and blocked with mercury, another heavy metal neurotoxin. There were devastating effects in all these situations. Users of lead dinnerware became mad as did users of lead based make-up, and cleaners of men's hats eventually became known as mad hatters. Most recently, evidence has been presented about the toxic effects of lead-based paint. Lead was added to paint to increase the whiteness. Of particular concern was the effect of lead-based paint on children. Children living in older homes would frequently peel off paint chips from the walls and consume the chips. There were long-term toxic effects of these childhood consumptions. Cadmium, another toxic heavy metal, which is used in making batteries, can have toxic effects in unforeseen ways. Cadmium can be discharged into a sewage system at the factory or can be discarded into the same system by consumers. It then contaminates the water that the sewage treatment plant discharges with possible subsequent toxic effects.

Individuals can wittingly poison their brains with substances they consume. Alcohol, nicotine, and cocaine, for example, can be toxic to the brain when consumed in excess. Death can result from a single night of binge drinking or from an extended period of heavy drinking. Alcoholics are often sent to "detox" treatment centers or critical care hospitals to remove toxic substances. After the detoxification process, recommendations are often made to alcoholics to enter a substance abuse center. The purpose here is to bring about behavior change. Subsequent to a stay in a substance abuse center the alcoholic may be referred to an AA (Alcoholics Anonymous) group to maintain behavior changes with the support of other group members.

Neurotoxins destroy brain tissue and symptoms of this destruction can include memory loss, changes in personality, difficulty concentrating, irritability, tremors, fatigue, seizures, depression, and dementia. Any one or any combination of the symptoms can affect communication. If the dementia stage is reached, communication is confused and detached from reality. Short of this advanced stage is the judgment by friends and family members who will often say that the sufferer is just not himself anymore. This judgment is based on both verbal and nonverbal cues.

# Psychopathologies

## Autism

The term, "autism," refers to a group of developmental disorders. The cause or causes of autism are unknown. In the mid-twentieth century parents were blamed. Parents with above average intelligence but who were emotionally cold were believed to be the ideal candidates to produce an autistic child. After the discovery of mirror neurons in the 1990s autism was believed to have been caused by the failure of the mirror neurons in the autistic child's brain. Even more recently Park (2011) reported that autism occurs more often in children who have an older sibling who is autistic. If there was an older sibling with autism, there was a 19 percent greater occurrence of autism in a younger sibling. A different finding on the cause of autism comes from Warner (2011). Warner reports on new marriage patterns that may have an impact on the increase in the number of autistic children. At present it is believed that at least one in one hundred and ten children is autistic whereas in 1980 the condition was believed to exist in only one out of 2,500 kids. Warner reports that there is an increase in the number of bright technical men marrying bright technical women. For example, Warner talks about the number of MIT marriages (both spouses were MIT graduates) that have produced an autistic child. The offspring were usually diagnosed with Asperger's, a mild form of autism. This view had some echoes of the earlier view that parents with similar intelligence and personalities fit the profile for autistic child parents. Finally, Sanders (2011a) reported on a number of studies supporting genetic factors involved in autism. Briefly, the researchers reviewed by Sanders believe that certain proteins known to be related to autism seem to be linked with over five hundred other proteins that may be involved in the emergence of autism. The research is preliminary and does not explain a number of mysteries.

Autistic children have serious communication problems particularly in the interpersonal setting. This magnifies the problems they have in social relationships. In the interpersonal setting the autistic child may have great difficulty establishing and maintaining eye contact with a conversational partner. If even a minimal relationship with another person is established, the autistic child will inevitably do or say something that will seem bizarre to the relationship partner causing stress or an end to the relationship. Alone or with others the autistic child will often exhibit repetitive vocal and/or bodily behaviors. Imagine, for example, an eight-year-old boy who spends his day repeatedly saying "a double decker car." With each repetition he increases the loudness of his voice and runs his fingers up the side of his body. He does this repeatedly until he seems to be exhausted. After "recovering," the boy begins the same routine once again. This occurs day after day.

## Schizophrenia

Schizophrenia is a serious psychopathology. Schizophrenics may hear voices, often voices of gloom and doom; can suffer distortions in thinking and perceptions of reality; and can have serious problems with interpersonal relationships. Attempts have been made to categorize schizophrenia. Five types have emerged: Paranoid schizophrenics feel that they are being persecuted; they also suffer from hallucinations. Disorganized schizophrenics exhibit confused and disorganized speech and inappropriate behaviors. Catatonic schizophrenics appear to be non-responsive to their surroundings and the people in those surroundings. Sometimes they will echo words said by others in their environments in parrot-like nonsensical fashion. Undifferentiated schizophrenia can include a number of the symptoms in the previous three categories. Finally, the residual schizophrenic exhibits fewer symptoms than exhibited at an earlier time. This often comes about as a result of therapeutic intervention, either and/or of talk or drug therapy. There are a number of anti-psychotic drugs that can ameliorate schizophrenic symptoms.

Schizophrenic speech has often been described as word salad. The words are just tossed together like the ingredients in a salad. A common response from a naïve listener to schizophrenic speech is that at first the schizophrenic seems to be producing a riveting narrative. But then there is the feeling that this narrative just doesn't make sense. It is going nowhere or it is an elegant word game. Paranoid schizophrenics often speak of someone or some agency implanting something in their brains that is talking to them or that is monitoring their behaviors. A sentence heard from a disorganized schizophrenic was as follows: "It's raining Saturday nights all over the place." Note that this sentence retains grammatical integrity, but it makes absolutely no sense.

Schizophrenia is seen as the result of both environmental and genetic factors. Sapolsky (2005, Lecture 15) presented one of the classic studies supporting this assertion. Using adoption records from Denmark, researchers determined the incidence of schizophrenia in adopted children who were placed in four different types of adoptive homes based on the presence or absence of schizophrenia in one or both of the adoptive parents or one or both of the biological parents. This yielded four possible outcomes:

- When one or both biological parents and one or both adoptive parents are schizophrenic, then 16% of the adoptive children are schizophrenic.
- When both biological parents are not schizophrenic and one or both of the adoptive parents are schizophrenic, then 3% of the adoptive children are schizophrenic.
- When one or both biological parents are schizophrenic and both adoptive parents are not schizophrenic, then 9% of the adoptive children are schizophrenic.
- When both biological parents are not schizophrenic and both adoptive parents are not schizophrenic, then only 1% of the adopted are schizophrenic.

This study supports the notion that the occurrence of schizophrenia has a strong genetic base, but its occurrence is not exclusively genetically based; environmental influences are present as well.

## Summary

All communication ultimately focuses upon a single person who generates, receives, and interprets messages. This chapter has reviewed the ways in which a number of pathologies can disrupt these communication roles. Often the initial emergence of a communication problem can serve as a symptom of more severe problems to come. Fortunately there a number of therapeutic interventions available that can be used to treat a variety of communication problems. Moreover, for more severe communication problems, some compensatory interventions are available and are being developed using a number of technological approaches. The following chapter will discuss these efforts.

# *Chapter 9*
# Augmenting Communication

Imagine that you wanted to communicate with a person who was literate but could not speak or write and whose only motoric response was movement of the left eyelid, and the person with whom you wanted to communicate was literate. What would you do? Let us suppose you wanted to know the person's first name. One of the first things you would do is to establish a code of "yes" and "no." Perhaps you could use one blink for "no" and two blinks for "yes." Then you might begin by asking if the first letter in the person's name was an "a" and wait for a yes or no response from the person. You might proceed letter by letter until you got a "yes" response from the person for a particular letter. Assume that the nonspeaking person is a woman and that the first letter of her name is "r." You could make some guesses as to her full name, but there probably are too many women's names that begin with "r" for you to make an accurate prediction of the woman's first name. If you scanned the alphabet a second time and found that the second letter of the woman's first name was "e," then your guesses might be more accurate. Is the woman's name, Regina, Rebecca, Rebe, Rene, or Rennie, for example? This form of communication has often been called the "20-questions technique."

What should be apparent from the above is that the process of determining just the first name of a person who cannot speak and has essentially no motor responses (only eye blinks) is a laborious and time consuming process. For this reason people who work with nonspeakers have devised a number of strategies to increase the efficiency of the process. In all cases, however, a reliable "yes" and "no" response must be established. For example, in scanning the letters of the alphabet one can ask is the letter in the first half of the alphabet or the second half? If one is going to communicate regularly with a nonspeaker, communication boards can be constructed. These boards can show the alphabet segmented into the upper half versus the lower half, and then subdivided yet again.

into the upper half of the upper half and the lower half of the upper half, for instance. It was such segmenting of the alphabet that allowed Jean-Dominique Bauby to write a novel: *The Diving Bell and the Butterfly: A Memoir of Life in Death.* Bauby was the editor-in-chief of *Elle* who suffered a brainstem stroke at age forty-three. Because of the stroke Bauby became locked-in, resulting in his ability to blink only his left eye. His brain was left intact so he could think and feel emotion, and he could see, hear, feel, taste, and smell. *The Diving Bell*, part of his book title, referred to his being locked into his body while the butterfly metaphor portrayed his freedom to think and imagine.

Other strategies can be used to increase the efficiency of communication (which is almost always defined in terms of speed). One can prepare posters or communication boards with certain themes posted on them: food themes, health themes, bathroom themes, family themes, and so on. Under each theme can be subthemes, for example, the health theme can refer to a number of symptoms. Nonspeakers can watch their assistants point to a number of options and signal "yes" when the correct option is indicated, and, of course, prediction strategies can be developed. Anyone who uses a search engine is familiar with this strategy. Type in only a few letters to begin a search and the search engine begins to make predictions as to what you are requesting. When an assistant is pointing to the letters on an alphabet poster or subsections thereof, the assistant can predict what the nonspeaker is trying to communicate. Context is very important here. Assistants should not make wild guesses when making predictions. This tends to make the nonspeaker feel powerless.

In all approaches to helping the nonspeaker, communication reliability of response is key. If a nonspeaker cannot reliably signal a "yes" and "no" response, communication is not possible. Speed of response (often expressed in characters per minute) is often sacrificed to achieve and maintain reliability. But speed of nonspeakers' communicative responses is extremely important. Conversation has a rhythm, and when this rhythm is broken conversation is often broken and terminated as well. One of the most common complaints made by nonspeakers is that people will not take the time to communicate with them.

It is not surprising that technology has been recruited to help nonspeaking individuals communicate. In fact, a whole field has developed around this effort. It is called the field of augmentative and alternative communication. Augmentative communication specialists often divide their efforts into those that are low tech and high tech. Low tech solutions would involve the creation of posters with an alphabet segmented into sections (upper half versus lower half, for instance), or a poster board with themes exhibited, for example. High tech solutions involve computers, particularly computers with voice synthesizers; these devices are often referred to as augmentative communication devices and are fabricated and sold as such. Access to an augmentative communication device is varied depending on the user's residual capacities. Some users may have body movements that are impaired but can be used to activate keys on a standard keyboard. Others can activate their devices by hitting a switch activated by any functional muscle. For example, a user may activate a switch by hitting it with a leg movement. Others activate switches by a sip and puff strategy, using a straw-like attachment which can activate a switch by either blowing or sucking on the straw. More sophisticated is typing out a message through eye movements. Here the user looks at the light-sensitive keys of a keyboard for a preset period of time and the key is activated. In many of these applications, the augmentative device will scan the keys on a screen keyboard automatically and the user will only be required to signal a "yes" response when the automatic scanner reaches the desired key. It is possible to adjust the speed of the automatic scanners to accommodate users' varied response rates.

What the computer-driven, augmentative communication devices have done for nonspeakers is to change them from being purely passive communicators to communicators who can initiate conversations and control the content of a conversation. This represents a major power shift. To prepare themselves for this new role, nonspeakers, with the help of their assistants, in some cases preload their augmentative devices with commonly expressed conversational utterances: what is your name? where do you live? how old are you? and so on. They will also preload their devices with answers to these same questions. The reason for the preloading is speed. The nonspeaker can trigger the release of an entire utterance with just one or two keystrokes, and then the augmentative device's speech synthesizer will retrieve that utterance and "speak" it. Utterances that are predictive in conversation are often put in a category called "quick talk." When you meet a co-worker in the morning the common conversational exchange includes, "How are you? Fine. How are you? Fine." These conversations are ritualistic, performed day after day. Other utterances in conversation are unique or novel and, therefore, cannot be stored beforehand. These utterances fall under the category of "exact talk.. When produced from "scratch" these utterances are typically generated very slowly by augmentative device users. Some sophisticated software for augmentative devices has been developed to increase the speed of generating exact talk messages.

With the advent of smart phones and tablets it is not surprising that someone would come up with an augmentative communication application. Leiber (2011) describes the work of Ajay Godhwani who built a text to speech app for his aunt's iPad. His aunt was diagnosed with amyotrophic lateral sclerosis four months earlier and lost the ability to speak. Godhwani first learned that two hundred words make up 80 percent of daily conversations. He used this information to construct a system which displays on the top of the user's computer screen the fifty most common words and phrases in English. The user can click on any one of these words or phrases at the top of the screen or can begin typing on the keyboard at the bottom of the screen. If the user begins to type, predictive text technology kicks in recommending words and phrases based on the first few letters typed. This can increase the speed of creating and "speaking" (with a speech synthesizer) an utterance. Godhwani's basic application is free but a premium version of it is being sold.

# Brain-Computer Interfaces (BCI)

The preceding paragraphs serve as an introduction to Brain-Computer Interfaces (BCIs). In BCIs electrical signals from the brain are captured and analyzed by a computer. The computer can then activate a prosthetic device such as an artificial arm for a quadriplegic, an environmental control such as the HVAC system in a room, a media device such as the television, or, for example, activate an augmentative speech device. An augmentative device is activated either by initiating a scan on a screen and composing a message choosing items from the screen, or by relaying the neural signals triggered when the device user imagines actually saying the message. Given the topic of this book the BCI discussion here will focus on augmenting the communication of nonspeakers. In adults the key causes of the brain damage that results in an individual's inability to use speech as a primary means of communication are congenital (e.g., cerebral palsy), degenerative (advanced amyotrophic lateral sclerosis, Huntington's disease, and Parkinson's), and trauma. Traumas can be external (gunshot and stabbing wounds) or internal (stroke). A brainstem stroke often results in the most severe paralysis of the body's motor system. Victims of a brainstem stroke are often labeled "locked in." People who are locked-in are essentially totally paralyzed, yet they can see, hear, feel, taste, and smell. Some can move only their eyelids (such as in Mr. Bauby's case described above). Others have no body movements whatsoever. Locked-In brain syndrome is usually caused by a brain stem injury or a brain stem stroke. Herculean BCI efforts are being made to help locked-in brain victims and other severely impaired individuals communicate. Two basic approaches are utilized in these efforts: non-invasive and invasive. In either case the initial goal is to establish a reliable binary or "yes" and "no" response.

## Non-Invasive BCI Approaches

Wolpaw et al. (2002) have published a comprehensive and extensive review of BCI research. The larger part of the review focused on non-invasive BCI approaches. The review is organized somewhat along geographical lines and a lead researcher from each geographical location is a coauthor on the review paper. The six non-invasive BCI approaches all use EEG monitors placed on the scalp. The monitors pick up the electrical activity in the brain and feed that activity to a computer. Recall that the brain is always active even in sleep so there is always "chatter" coming from the brain. The computer is scanning this continuous input or noise and is looking for a blip or signal that is unique and relates to some thought on the part of the user. For example, if a speaker broadcasts the words, "dog, tree, rock, thought, and DVD," and the user is to respond to "rock," the computer will look for any neural signal that occurs almost simultaneously subsequent to the broadcast of the word, "rock." This routine follows the classic signal to noise analysis. In common parlance, for example, even a person with normal hearing may be unable to understand the speech (the signal) of a companion in a very noisy club.

The BCI approaches discussed by Wolpaw et al. (2002) are the visual evoked potential approach, the slow cortical approach, the P300 approach, the mu and beta rhythm approach, the Wadsworth approach, and the Graz approach. The visual evoked potential approach rests on the users' eye movements. The user looks at an 8 x 8 matrix with alphabetic letters and other symbols. The EEG pickups on the back of the user's head tracks the direction in which the user is looking and the specific letter at which the user is looking. That letter is now posted on the screen and the user then looks at the next desired letter, then the next desired letter, and so on. A user trained and experienced in using this technique, can generate ten to twelve words (word length = 5 letters) per minute. A user with uncontrollable head and neck movements can contaminate this procedure and make it unusable. It is important to note that this technique is distinct from eye tracking.

The slow cortical potential approach is the approach that University of Tubingen researchers have been exploring for years. The lead researcher is Niels Birbaumer. Birbaumer noted that in the EEG display the lowest frequency shows low voltage changes in the cortex. These signals are called slow cortical potentials (SCPs). The Birbaumer team found that people can learn to control their SCPs. For example, when a subject thought about anticipation of movement but not actual movement there would be a positive SCP. On the other hand, when a subject thought about release or actual movement there was a negative SCP. A good image for the subject was to imagine being a runner in a starting block (anticipation) and when the starter pistol was fired (release), a negative SCP was provoked. The Tubingen discovery led to the development of the thought translation device (TTD). The SCP recordings provide the user of the TTD a binary choice or a "yes' or "no" ability and, therefore, a user can communicate either in printing or speaking through a voice synthesizer by indicating whether the desired letter is in the upper or lower half of the alphabet, the upper half of the upper half or the lower half of the upper half, and so on. The system permits users to produce up to three letters per minute. Parker (2003) has written a detailed case study which portrays Birbaumer working with a locked-in brain patient.

When significant auditory, visual, or tactual stimuli are injected into the stream of routine or mundane stimuli received by the brain, there is often an EEG peak evoked at about 300 milliseconds. This is the basis of the P300 approach to BCI. In operation, a 6 x 6 matrix is presented to the user on the computer screen. There is a flashing of each row and then each column of the matrix in sequence. This is done twelve times. The computer analyzes the user's brain waves to determine on which row and which column a P300 response occurs. In a matrix, a given row and a given column identifies a specific cell in the matrix. For example, if a user wants the letter "c" to be the first letter in a message, the user will predictably exhibit a P300 response on Row 1 and Column 3. There are, of course, only twenty-six letters in the English alphabet and thirty-six cells in the matrix so the remaining cells can be used for "backspacing," "return," and so on. The advantage of the P300 system is that little or no training is involved. Use of the P300 method for communication would permit a user to produce approximately one word per minute.

The Wadsworth BCI method utilizes mu or beta-rhythm amplitudes to control the bi-directional movements of a cursor. Here again binary control is achieved. Users learn to move the cursor up or down. When beginning to learn this technique, users will often imagine up and down movements of some kind, for example, hand or whole body movements. When a reliable binary response is established the Wadsworth BCI can be used to select letters to communicate or to use environmental controls. Learning to use this method takes training, but at the end of this training users can attain accuracies greater than 90 percent for answers to "yes" and "no" questions.

The Graz BCI, named after the home of the research, the University of Graz, is similar to the Wadsworth BCI in that it uses mu and beta rhythms. The program has users imagine different and simple motor actions, moving the right or left hand, for example. This imagery establishes a binary response which, after training, is reflected in the EEG information presented to the computer for analyses. Although the Graz data can be utilized to communicate, the research seems to be aimed primarily at the controlling of a prosthetic device. The system is rather easy to learn. About 90 percent of test subjects can learn the system and can achieve 90 percent accuracy.

Pushing the envelope a bit further has been research conducted at the University of Southampton in 2009 which demonstrated that it is possible for one person to communicate with another through the power of thought. The research was conducted by Christopher James. The basic setup was to have Person A, the transmitter, generate a series of binary digits by imagining moving his right arm or left arm. The thoughts were picked up by an EEG recording and computer analyzed. The signals generated by Person A's thoughts were then amplified and sent to a second computer which flashed the binary signals on an LED. Person B, the recipient, saw the flashes and EEG signals from Person B's brain showing that Person B had accurately received the messages from Person A's brain. The messages, zeroes or ones, were of course rather primitive, but the research opens a new path for research, nonetheless.

## Invasive BCI Approaches

Invasive BCI approaches require brain surgery wherein a device is implanted in the brain of the recipient. The implanted device has a number of very thin wires that are inserted into neurons in the recipient's motor cortex. The device picks up electrical messages from the implanted wires and transmits these messages to a computer which once again tries to separate the signal from the noise. Some of the devices strive to pick up binary responses, "yes" or "no," push or pull, move right or left, or move up or down. An implant is a foreign body and the body's normal tendency is to wall off a foreign body. Thus scar tissue can grow around the wires, and the wires can corrode, and, as a result, can lose their transmission capacities over time. Therefore, at the beginning of invasive BCI research a number of bio-engineering problems had to be solved in non-humans before human trials could be conducted. Human trials have involved the movement of robotic limbs, computer operation, playing computer games, environmental controls, and human communication. The focus here is on the last item. Two major invasive BCI research efforts are underway. Both efforts have been incorporated. The efforts are BrainGate Incorporated (Hochberg et al. 2011, Simeral et al. 2011, Hochberg 2008, and Hochberg et al. 2006), and Neural Signals Incorporated (Kennedy and Bakay 1998, Brown 2008).

BrainGate recently celebrated the one thousandth anniversary of one of its implant subjects. The subject, a woman in her late fifties, was a quadriplegic as a result of a brainstem stroke, was unable to speak or use any of her limbs. She was implanted in 2005. One thousand days after the implant the woman was still able to operate a computer cursor in spite of the fact that the implant exhibited poorer signal quality at one thousand days than it did at six months. The slow decay of the signal over time has been attributed to engineering, mechanical, and procedural issues, and the BrainGate group hopes that eventually their implants will provide decades of useful signals. An earlier implant in a twenty-five-year-old male, who was stabbed in the neck and paralyzed from the neck down, showed that an implanted person could talk (this patient was able to speak) while at the same time accurately moving a cursor on a computer screen, opening emails, playing computer games, and adjusting the controls on a television just by thinking of doing these things. This means that thought produced movements do not require total concentration. What both of these cases have demonstrated is that surgically implanting a sensor in the motor cortex is not necessary immediately post trauma. Brain activity in the motor area seems to persist long after spinal cord or brain stem injury. The sensor that BrainGate implants is 4mm-square and has one hundred sensors recording the electrical activity in the motor cortex. These electronic messages are then transmitted to a computer for message extraction and subsequent action, cursor movement, environmental control, prosthetic control, or augmentative communication, for instance.

Neural Signals Incorporated is working in the same research areas as BrainGate. Its chief researcher, Philip Kennedy, performed the first implant surgery on a human, a woman in the terminal stages of amyotrophic lateral sclerosis also known as ALS and Lou Gehrig's disease. In two months the woman learned to do several operations on a computer through thought. Several years later Kennedy implanted a neural prosthesis in a fifty-three-year-old male dry wall contractor who was a stroke victim. The man learned to move a computer cursor, select computer icons, and spell words so that speech could be generated through an augmentative communication device. Kennedy has developed a number of procedures for keeping the neural prostheses in place in the brains of his subjects. Brain tissue has a tendency to move as the head is moved. This disrupts the signal flow from the transplanted devices. The extreme example of this movement is that that occurs in closed head injury where brain tissue actually "bounces" around in the skull causing injury. One solution is to inject wax around the transplant to help hold it in place. Another is to have the transplant wires placed in cones coated with a chemical that promotes neuron growth. Here neurons actually grow into and around the transplant. In addition, the implant can be stabilized by anchoring it to the skull with screws.

One of Kennedy's current research efforts focuses on communication. He has said that his team learned that if a BCI patient can produce a binary response by thinking then an augmentative communication device can be utilized to produce speech by having the device scan the alphabet or parts thereof and the device user can signal "yes" or "no" to accept a certain letter or segment of the alphabet. This is a very time consuming process. Kennedy wants to have a subject think of the sounds of a language that are in the word that the subject wants to say. For example, the word "bet" has three sounds: /b/, /e/, and /t/. If a subject can think of those three sounds and a computer can pick those sounds out from all the other signals that are simultaneously emanating from the brain, then the speed of speech production can be markedly accelerated. This is a daunting task. The greatest problem is having a computer program that can detect the speech signals from all the other signals that are emanating from the subject's brain. Aiding Kennedy in this effort was Frank Guenther (2006) who developed the sophisticated computer program. The first subject to try this system was able to produce three to five vowel sounds after a number of tries. The computer program driving this project provides feedback to the subject permitting the subject to alter his thoughts on subsequent tries. Guenther compared the learning in this situation as unconscious learning comparable to a person learning to shoot baskets, for example. As you keep trying you just get better; and so it is with this speech program. Kennedy believes that this trial (thinking and thereby triggering the actual production of speech sounds) to be a first step to having a subject speak in fluent sentences.

## Summary of BCI Approaches

It is important to note that BCI approaches, whether non-invasive or invasive, are still in the experimental stage. There is no clinical deployment of BCI devices in either category. Most of the non-invasive research uses normal subjects. Even with these non-involved subjects sequential learning trials must be conducted to attain acceptable accuracy on very basic tasks. Moreover there are wide individual differences and sometimes there is decay in accuracy over time. The invasive methods use severely impaired individuals as subjects, trauma victims, or sufferers of several degenerative diseases because there is always risk associated with brain surgery such as infection. Surgical implants designed to operate mechanical arms have always been previously tested with non-humans to support the efficacy of any intervention with humans. But, of course, non-humans do not talk so any explorations in the speech and language areas have to involve humans. Non-humans can be used to test the viability of an implant, its stabilization, and its life span, but where speech and language are involved, the subjects must be human. And these human subjects are already suffering from the trauma or disease that made them eligible as subjects. For this reason many subjects tire quickly and training sessions have to short and spaced.

Progress in the BCI area will proceed. But it will be a long time before an "off the shelf" device can be prescribed and used routinely. Required is the mobilization of a cadre of professionals from many disciplines, bio-engineers, surgeons, computer programmers, psychologists, special educators, speech and language specialists, and seating and positioning specialists (usually occupational or physical therapists), for example. The effort has to be cooperative and it is labor intensive. To date, virtually all BCI subjects have been literate. The non-reader presents additional challenges. Literacy can be a problem for young pre-school cerebral palsied children and illiterate adults who suffer from a physical pathology. In the field of augmentative communication, when the learner is illiterate, images called icons are often used. Thus the picture of an eye may be used to represent the first person singular. The speech work of Kennedy may ultimately be a solution for some of the communication problems of the non-readers.

One advantage that locked-in individuals, spinal cord individuals, and degenerative disease-afflicted individuals have is that their senses and cognitive skills remain intact, although perhaps dimmed in individuals with degenerative diseases. This means that they can always receive feedback for their efforts. For example, locked-in individuals may not be able to move a single muscle, but for the most part they can hear, see, feel (for spinal cord injuries, this would depend on the level of the injury), taste, and smell (sadly a fact unknown to many family members and even professionals). For BCI efforts not related to communication, feedback becomes a problem. For example, a quadriplegic operates a prosthetic arm and reaches for a cup of coffee. The quadriplegic cannot feel the temperature of the coffee, nor can the fragileness of the paper or Styrofoam cup be determined. This then brings up the problem of sensation. Can there be signals sent from the environment to mimic sensation, in this case, tactual and thermal. Working on this problem is Miguel Nicolelis (2011), who believes that signals can be sent not only from the brain but to the brain as well. He is particularly interested in having users of thought-controlled arms or legs (prosthetic or real) touch and feel things as they interact with the world. Is that coffee too hot or is that cup too fragile for the strong grasp of my robotic hand? This information is critical and Nicolelis wants to provide it. Of course some people recoil at the idea of sending information directly into the brain. Could this eventually be used to brainwash individuals? The notion brings to mind the old swords to plowshares idea and vice versa. Almost any technology can be used for good or evil. A psychiatrist might implant some ideas directly into a patient's brain that would relieve a patient's suffering, or on the other hand, an intelligence officer might implant some ideas directly into an agent's mind that would cause the agent to commit some terrorist act.

## Summary

Many non-BCI approaches are already being prescribed for nonspeaking individuals both literate and illiterate. Use of these devices has helped many nonspeakers communicate in ways that were impossible before an augmentative device was prescribed for them. The devices have been particularly helpful for the transmission of speech and language messages. Still many nonspeaker have communication problems. Prominent among these is the rate of speech output. Many listeners will simply not wait for slow responses. Nonverbally many nonspeakers with neuromuscular problems are stigmatized from the outset because of their seemingly inappropriate movements and postures. Many nonspeakers, therefore, need sympathetic conversational partners.

BCI approaches are still in the experimental stage. As mentioned above, the noninvasive approach research efforts use normal subjects while the invasive research efforts use subjects with serious pathologies. Possibly BCI research will advance much more rapidly when devices are developed that have wide application. For example, when a device is developed that allows a driver to operate an automobile through thought-generated messages, more non-invasive devices for the handicapped will also become available.

# *Part III*
# Introduction

In Part 3 there is a looking back and a looking forward. Chapter 10 takes a critical look at the methodologies used in brain-based communication research. Chapter 11 looks at the ethical issues that have arisen as a result of recent brain research. Finally, chapter 12 looks to the future and what we may expect. Again one can take the ancient swords versus plowshares view. Many of the results of future research can be used for good or evil. The question is which use will be implemented?

# *Chapter 10*
# Methodological Issues

## Major Issues

Three major methodological issues cloud neuroscience research in general and neuro-communication research specifically. The first two issues deal with the principal technology used in behavioral neuroscience: fMRI. The third issue deals with the paradigm shift in neuroscience which may require a re-writing of much of the literature.

### General Comments on fMRI

Bandettini (2007) has provided a general review of fMRI. Bandettini has noted that in 1992 there were no published papers where fMRI was used. In the year 2000 there were almost one thousand papers published using fMRI, and in the year 2005, 2,500 papers using fMRI were published. These numbers show the rapid growth in the use of fMRI in neuroscience studies. With this rapid growth it might be expected that many studies using fMRI would be poorly planned, executed, and analyzed. And this has been the case, according to Bandettini. Moreover, Bandettini has noted that many of the studies that were completed reflected a "frantic rush to pick the scientific 'low-hanging fruit.'" On the positive side, according to Bandettini, the fMRI studies have provided some unique insights on how the human brain is organized, and has recruited a larger number of researchers into the neuroscience area.

Bandettini has noted three major weaknesses with fMRI: temporal resolution, spatial resolution, and interpretation. Regarding the temporal problem, there can be a latency of up to four or five seconds between the beginning of brain activity and the fMRI reflection of that activity. On the surface it would appear that fMRI provides fine spatial resolution. Bandettini cautions, however, that fine spatial resolutions may not translate into accurate resolutions. Finally, Bandettini notes that the interpretations made of fMRI signals are based upon the relationship between neuronal activity and hemodynamic signal changes. This relationship appears to be valid, but more studies are needed to move the relationship to one of causality from one of correlation.

## Methodological Issues in Neuromarketing

Touhami et al. (2011) list some of the methodological issues in neuromarketing. They noted that research protocols in neuromarketing are long and difficult to elaborate, that the number of subjects is generally weak, that the results neuromarketing research give are seldom important and clear enough to allow a significant statistical treatment, that some techniques used in neuromarketing (such as fMRI) can be unpleasant or uncomfortable, that neuromarketing research is expensive (ten subjects may cost 50,000 dollars), and finally, neuromarketing research requires many consent procedures since subjects are undergoing brain imaging. The approval of an ethics committee is also sometimes required.

## Brain Scan Flaws

Shermer (2008) has listed five major flaws with fMRI scans.

- First, he notes that the fMRI scanner is an unnatural environment for cognition and that about 20% of subjects become claustrophobic and cannot finish the experiment for which they were recruited. This presents a problem of subject bias. There cannot be a true random sample. Furthermore, subjects' mobility is severely restricted. Heads must be locked firmly in place. Visual stimuli are presented through the small lenses of goggles.
- Second, Shermer notes that scans are indirect measures of brain activity. When the brain is active it uses oxygen. Oxygenated blood is pulled into the activated neurons. The oxygenated blood is rich in iron and the magnet in the fMRI scanner is sensitive to iron and thus active areas in the brain are detected. But at most the relationship between the stimulus used and the blood flow response is correlational, not causal. The brain is active in the absence of any external stimulation.
- Third, Shermer notes that colors can exaggerate the effects of external stimuli on the brain. When there is activity in a certain area the brain scan area is usually presented in red, signaling oxygenated blood flow. But how are the boundaries of red decided upon? It is somewhat arbitrary. Actually brain scan data could be presented in numbers, but colors are more dramatic.

- Fourth, brain images are statistical compilations, notes Shermer. A brain scan session can range from minutes to two hours. During this time hundreds to thousands of images are generated. The data is then tweaked somewhat to compensate for head movements and differences in head sizes, for example. The images are then lined up, averages are taken, and colored images are produced.
- Lastly, Shermer notes that brain areas activate for various reasons. Just because a brain area lights up in response to a given stimulus presentation does not mean that that is the sole function of this area. The amygdala lights up when a fear-generating stimulus is presented. But that same structure is activated also when a subject feels positive emotions or is aroused. Probably no brain area responds exclusively to a single stimulus. In addition, when a stimulus is presented to a subject in an fMRI scanner, much of the brain is active just processing the stimulus. A visual stimulus, presented to the subject through goggles, for example, activates the occipital lobe and other regions of the subject's brain which are not the primary interest of the experimenter.

## Questionable Statistical Analyses

In 2008, an article entitled "Voodoo Correlations in Social Neuroscience" was accepted for publication and was "In Press" for the journal, *Perspectives on Psychological Science.* The title drew an inordinate level of attention from both scholars and the popular press. Begley, for example, wrote an article entitled "Of Voodoo and the Brain" for *Newsweek* (Feb. 9, 2009). The article by Vul et al. eventually appeared in 2009 with the new title, "Puzzlingly High Correlations in fMRI Studies of Emotion, Personality, and Social Cognition." The authors of the article noted that published research reporting correlations between scores on social-psychological instruments and fMRI results found inordinately high correlations between the variables. The authors reviewed fifty-five articles reporting high correlations and surveyed the authors of the reviewed articles asking the authors for details on how the correlations were computed. Three major criticisms emerged from the Vul et al. article: problems with reliability of the instruments (both hard and soft), problems with defining the fMRI variable, and problems with multiple comparisons.

### *Reliability Issues*

Vul et al. noted that psychometric practitioners have found that the strength of a correlation between two measures or variables is dependent not only on the strength of the relationship between the two variables but also on the reliability of the two variables. The reliabilities of the measures or variables, x and y, determine the upper limits on the correlations that can be obtained between x and y variables. The test-retest reliability of most social-psychological measures hovers around +0.7 to +0.8. The reliability of fMRI measures seems to hover around +0.7. Given these numbers the highest expected correlation between a social-psychological variable and an fMRI variable would be, according to the psychometric literature, 0.74. Recent studies correlating social-psychological variables with fMRI variables, according to Vul et al., routinely have exceeded this ceiling.

## *The fMRI Variable*

A functional magnetic resonance image shows the level of deoxygenated hemoglobin in the blood in a particular region of the brain. Many measurements are made in cube-shaped regions. The cubes are called voxels for volumetric pixels. Voxel volumes can range between one cubic millimeter and 125 cubic millimeters. Given this variation each functional image contains between 40 thousand and 500 thousand voxels. Typically a new functional image is acquired every two to three seconds. The question raised by Vul et al. was "How was the set of voxels selected?" The authors concluded that, in general, the voxels that were selected to be included in the correlation computations were those that correlated highly with the score on the social-psychological instrument. Authors need to provide more detail about extracting and analyzing fMRI data. Consider the following description of image analysis provided by Noordzij and colleagues (2009).Vul and his colleagues might not agree with everything stated in the analysis, but they would appreciate the description, nonetheless.

> Functional data were pre-processed and analyzed with SPM2 (Statistical Parametric Mapping). The first four volumes of each participant's time series were discarded to allow for T1 equilibration. The image timeseries were spatially realigned using a sinc interpolation algorithm that estimates rigid body transformations (translations, rotations) by minimizing head-movements between each image and the reference image. The timeseries for each voxel was realigned temporarily to acquisition of the middle slice. Subsequently, images were normalized onto a custom Montreal Neurological Institute-aligned EPI template (based on 28 male brains acquired on the Siemens Trio at the Donders Centre) using both linear and nonlinear transformations and resampled at an isotropic voxel size of two mm. Finally, the normalized images spatially smoothed using an isotopic eight mm full-width-at-half-of-maximum Gaussian kernel. Each participant's structural image was spatially coregistered to the mean of the functional images and spatially normalized by using the same transformation matrix applied to the functional images. The fMRI timeseries were analyzed using an event-related approach in the context of the General Linear Model (GLM). (p. 8)

The above quote is presented to show the complex operations executed in preparing fMRI data for analysis. It has been presented to give the reminder that many operations underlie the colorful pictures that we are accustomed to seeing in fMRI reports. The color red, for example, is a choice since it reminds the viewer of oxygenated blood. It could be any color or even numbers for that matter.

## *Multiple Comparisons*

Imagine that you have a random number generator that can generate random numbers between zero and one hundred. You now generate one hundred pairs of random numbers and compute a correlation between the pairs. You repeat this process one thousand times. By chance alone we would expect to find a number of the correlations to be positive and statistically significant even though with random number pairs we would expect, again by chance alone, to hover around zero. The message of this imaginary scenario is that with repeated comparisons we would expect by chance alone that some of the correlations would be positive and statistically significant. In statistics, of course, any result obtained can have occurred by chance alone. But some of these results are of such high magnitude that the researcher declares the result to be real even though the result could have occurred by chance alone. Behavioral science researchers often declare that if some result could have occurred by chance alone 5 percent of the time or less, it is a real outcome or a statistically significant result.

If a researcher computes a correlation coefficient and states that if the coefficient found is one that would occur by chance alone 5 percent of the time or less, the researcher will declare the result to be significant. Then if the researcher computes a second correlation coefficient there is another chance that the second correlation will be one that would occur 5 percent or less by chance alone, but according to the researcher's previous declaration, the correlation is declared to be real and statistically significant. But wait. If there was a 5 percent chance of getting a significant correlation in Trial One and a 5 percent chance in Trial Two, then there is a 10 percent chance of getting a significant result. As the number of trials or the number of correlations increases, the chance of getting significance increases dramatically, even with random number pairs.

In many of the studies reviewed by Vul et al., an excessive number of correlations between the social-psychological variable and the fMRI variable were computed. In many cases thousands of correlations were computed. Given this, it would be expected that even very high correlations could occur by chance alone. What troubled Vul et al. was that some researchers selected only those correlations that were high and statistically significant.

All need not be lost, however, according to Vul and his colleagues. The authors of the studies reviewed and critiqued by his group still have their raw data and they are, therefore, able to re-analyze their data using more rigorous standards. Reporting their procedures will promote a key principle of science, published research studies should describe their methods to the extent that the studies can be replicated.

## The Broader View on Theory and Methodology

In 1962 Thomas Kuhn published *The Structure of Scientific Revolutions*. The basic thesis of the book is that science does not progress in an evolutionary way, but in a revolutionary way. These revolutions or disruptions were referred to by Kuhn as *paradigm shifts*. When there are paradigm shifts, the shifts are often met with disdain, then doubt, and finally acceptance. The neuroscientist Dale Purves (2010) appears to be calling for a new paradigm in neuroscience when he asks why we remain so ignorant of brain functions. Purves seems to echo the sentiment expressed by Paul Allen at the beginning of chapter 2 in this book.

In answering his question, Purves suggests that there is an absence of a guiding principle or principles that would help us understand the neural underpinnings of perceptual, behavioral, and cognitive phenomenology in a more general way. He notes that other areas of biology have benefitted from having such guiding principles: cell theory, the theory of evolution, and genetic theory, for example. Because of the complexity of the brain and the nervous system the absence of a general guiding principle is understandable. If the history of science is any guide, a general principle is bound to emerge, according to Purves.

## Summary

Reports criticizing methods used in neuroscience research show that the scientific method is working. A hallmark of the method is that it is self-correcting and that errors will eventually be corrected. Instead of despair at finding errors and weaknesses in published studies, readers should relish the fact that the scientific method is indeed working.

# *Chapter 11*
# Ethical Issues

In mid-twentieth century, there began a revolution in biology and medicine. The discovery of the structure of the DNA molecule ushered in a new era in biology. This was followed by the proposal to use stem cells to treat a number of diseases and conditions. Comparisons were made with the early part of the twentieth century which was deemed to be the age of physics; there was the theory of relativity, and there were discoveries in nuclear physics. With each discovery in each age ethical issues emerged. Should the atomic bomb have been used in Nagasaki and Hiroshima? Should individuals' genetic histories and their genomes be revealed to them as part of their medical histories? Should it be legal to treat a patient with stem cells from an aborted fetus? While there was no moniker for the ethical issues that emerged in physics, there was one in biology, specifically, bioethics.

Now we are in the age of the brain. The number of scientific studies on the brain has increased exponentially, and the number of citations to the brain in popular media has similarly increased. Not surprisingly, ethical issues emerge in tow, and not surprisingly, there is now an area of study called neuroethics. Three major neuroethical issues will be discussed in this chapter. The first brings to the surface an old philosophical question, the question of determinism versus free will. The second relates to the end of life. What is death? And, who determines when death should occur? And the third reviews the ethics of pre-emptive actions.

# Determinism versus Free Will (My Brain Made Me Do It)

Chapter 6 in Michael Gazzaniga's book, *The Ethical Brain* (2005), is titled, "My Brain Made Me Do It." In this chapter, Gazzaniga tackles the determinism-free will question. Appropriately, he discusses the issue within the context of the American courtroom.

> Against this real backdrop of what life is like in the American courthouse, a new wrinkle is appearing in the form of the perennial question, Do we as a species have "free will"? Did the defendant carry out the horrible crime freely and by choice, or was it inevitable because of the nature of his brain and his past experiences? As with so many issues where modern scientific thinking confronts everyday realities, the people in the jury box are not rushing to embrace this one. Yet it is my contention that even those tough jurors will have no choice, because some day the issue will dominate the entire legal system. (p. 88)

Gazzaniga notes that the determinism versus free will question is not a new one.

> These dilemmas have been haunting philosophers for decades. But with the advent of brain imaging, neuroscientists are exploring these questions, and, increasingly, the legal world is demanding questions. Defense lawyers are looking for that one pixel in their client's brain scan that shows an abnormality, a predisposition to crime or a malfunction in normal inhibitory networks, thereby allowing for the argument, "Harry didn't do it. His brain did it. Harry is not responsible for his actions." (p. 89)

On the determinism-free will issue Gazzaniga acknowledges that some individuals with anti-social personality disorders have less gray matter than normal subjects. At the same time this does not necessarily mean these individuals can not act responsibly. Gazzaniga concludes that neuroscience will never find, nor will neuroscientists ever make declarations of responsibility or irresponsibility. The issue of responsibility is a social choice or social construct, according to Gazzaniga.

What Gazzaniga used as a chapter title, Sternberg used as a book title: *My Brain Made Me Do It* (2010). In the book, Sternberg also examines the determinism-free will issue. He sets the stage for his examination of the issue by describing a pizza store robbery in Atlanta, Georgia, in 1991. A man walks into the pizza store holding a semiautomatic pistol in his hand. The man points the gun at the single employee in the store and asks the employee to open the cash register. The employee complies and the robber empties the cash register. The employee sits in the corner trembling. The robber now tells the employee to get on his knees, moves behind the employee and shoots him in the head, killing him. The robber is arrested and tried. The defense team knew that its task would be a difficult one particularly since the evidence against the robber/killer was so overwhelming and because the defendant showed no remorse. In fact, the robber/killer showed disdain for his victim because the victim showed so much fear.

The defense team had a broad battery of tests administered to the robber/killer. Diagnostic tests for schizophrenia, bipolar disorder, and Alzheimer's, for example, were negative. Only one test showed a positive result. The robber/killer's brain had a deficiency of an enzyme that is responsible for breaking down serotonin and several other neurotransmitters. Eureka! The defendant's brain made him do it. The defendant was not morally responsible for his actions, said the defense team. In this case the jury did not agree; the defendant was found guilty and executed.

Sternberg uses the pizza store robbery as the backdrop to make his arguments. He notes that the robber had the freedom to decide to rob or not rob the pizza store, and that he could have worn a mask if he feared that the store employee would have later identified him or called the police and provided a description of him. Since the robber had a criminal record he was aware of the consequences of criminal behavior. Sternberg claims that the robber was a moral agent who planned his actions, and, across the board, knew the consequences of his actions.

Coming down hard on the determinism side of the issue are Dr. Kent Kiel and Danalynn Recer (Seabrook 2008, Toobin 2011, respectively). Kiel works with psychopaths at the Western New Mexico Correctional Facility. Using fMRI scans Kiel has concluded that there are differences in the brains of psychopaths and non-psychopaths. Psychopathy is caused by a brain defect, according to Kiel. Specifically, the defect is located in the limbic system, a network of brain structures stretching from the prefrontal orbital cortex to the posterior cingulated gyrus. These structures are involved in processing emotion, exercising inhibition, and attention control. The inductive leap, then, is that psychopathy is a mental disease and, thus, psychopaths are not responsible for their horrific crimes. Kiel believes that psychopaths are discriminated against because they have been portrayed as predators. In comparison, the schizophrenics get all the sympathy and attention because they are seen as victims.

Even though Kiel has compared the brains of the psychopaths with other members of the prison population, it is too soon to come to the conclusion that psychopaths are just marching to a different brain drummer. There may be others in the population who have the same brain anomalies but do not commit horrific crimes. On the other hand, if Kiel is correct that psychopathy is brain based, then society has to decide how to deal with the problem. Should every individual be brain scanned, and, if so, at what age should they first be scanned? If a brain anomaly is found, should the individual be incarcerated, be given preventive therapy, or should some other course of action be taken? Certainly if psychopaths were to be removed from society, many lives would be saved.

Danalynn Recer is a defense attorney practicing in Houston, Texas. For many years Houston had the record of being the death penalty capital of the world. Recer may have had a part in changing that. From 1972 to 1976 executions were put on hold in the United States while the Supreme Court considered the issue. In 1976 the court directed that executions could resume and dictated a two-phase structure: a guilt phase and a penalty phase. In the first phase the determination is made that the defendant is innocent or guilty, while in the second phase, the penalty is determined. In first degree murder cases it is usually death or a life sentence without parole. Recer has made her mark by arguing for death penalty defendants in the penalty phase. Her arguments describe the backgrounds of the defendants. She identifies the precipitating factors that contributed to the defendants' actions. She will bring to the stand witnesses who will testify about childhood abuse, abandonment, and loneliness, for example, that the defendant experienced. She will even bring to the stand victims of the defendant who now forgive the defendant. And, she will argue that brain injury was a contributing, if not causal, factor of the crime. Her arguments are designed to provoke both sympathy and empathy in the jurors.

As a case in point, Recer argued that a defendant in a murder trial not be executed but be given a life sentence without parole. The defendant had shot a police officer when he was under arrest and handcuffed. The defendant managed to pull a handgun out of his pants even when handcuffed and shot and killed the officer. It was learned that when the defendant was six years old he fell from a roof, was knocked unconscious, and had to be hospitalized. After this episode the defendant suffered from brain seizures. This led Recer to have the defendant brain scanned and as a result of the scan Recer built a defense on insanity. The murder of the police officer was a freak circumstance, not part of a pattern in the defendant's life, argued Recer. Recer also intertwined the fact of the defendant's alcoholism with his brain damage. The defendant received a life sentence without the opportunity of parole. Recer saved him from execution. And Recer said at the end of the trial that the defendant will now die when God decides and not when man decides. It was a victory for Recer and for determinism. The defendant's brain made him do it.

On August 1, 1966, Charles Whitman went to the top of the clock tower at the University of Texas at Austin carrying an arsenal of weapons. On his way to the observation deck of the tower he killed a receptionist with the butt of one of his rifles. He also shot a group of tourists at point-blank range. Then he proceeded to the observation deck and shot at random at the pedestrians walking on the street below. His toll for the day was thirteen people killed and thirty-two injured, one of whom died later. Whitman's assault was stopped after the police who climbed the tower shot and killed him. The grim nature of the story, however, was not over. When the police investigated Whitman's home for any clues to account for his behavior, they found the body of his brutally stabbed wife and later the body of his mother whom he also killed. What was also found were several notes typed and written by Whitman. In the notes he stated that he didn't understand himself because he was going to kill his wife even though he loved her dearly. He clearly did not know what was happening to him and asked that when he died an autopsy be done to see if something was wrong with his brain.

That autopsy was done and it was discovered that Whitman had a small brain tumor growing beneath the thalamus, impinging on the hypothalamus and the amygdala. The amygdala is very much involved in emotion and social disturbances.

In the year 2000, the sexual preferences of a normal man, we'll call Jack, began to change. What was once a conservative, married stepfather now became a man with an obsessive interest in child pornography. At home, Jack hoarded printed child pornography, he cruised child pornography websites, and he visited massage parlors, soliciting various sexual acts. He began making sexual advances to his prepubescent stepdaughter. At this point his wife responded and, with a court order, had her husband removed from the house. Complaints by Jack of persistent and worsening headaches resulted in a brain scan being performed for him. The scan revealed a massive tumor in the frontal-temporal cortex. Pathology in the frontal-temporal cortex area often results in an inability to follow norms and a tendency to give in to a variety of hidden impulses like shoplifting, for example. When Jack's tumor was removed in two separate operations, his behavior returned to normal.

A woman we will call Jill began taking a new drug to alleviate the symptoms of Parkinson's disease. Shortly thereafter Jill became an obsessive gambler, a compulsive eater, and a compulsive drinker. She amassed large gambling debts and became dysfunctional in social relations. The drug that Jill was taking was designed to mimic the effect of the neurotransmitter, dopamine. Dopamine is responsible for muscular movement, but it is also involved in rewards obtained in gambling, eating, drinking, and so on. Fortunately for Jill a change in dosage of the drug eliminated her compulsive behaviors while at the same time controlling her Parkinson's symptoms.

The anecdotes of Whitman, Jack and Jill have been adapted from neuroscientist David Eagleman's (2011a) essay entitled "The Brain On Trial." Eagleman believes that we are a product of our genes and our environment. These two factors are very deterministic. We cannot assume that people have the same capacity to make sound choices because their brains are vastly different, according to Eagleman. As brain science progresses, the role and existence of free will diminishes. Eagleman says:

> Free will *may* exist (it may simply be beyond our current science), but one thing seems clear: if free will *does* exist, it has little room in which to operate. It can at best be a small factor riding on top of vast neural networks shaped by genes and environment. In fact, free will may end up being so small that we eventually think about bad decision-making in the same way we think about any physical process, such as diabetes or lung disease. (p. 118)

Eagleman, of course, realizes where his argument is taking him: to the legal system. Can anyone ever be guilty of anything? Eagleman proposes a forward-looking approach to sentencing. If a suspect is found guilty, for example, the sentencing should be based on scientific principles. Brain scans would be routine, and as time goes by, these scans will become more and more sophisticated. An "eye for an eye" approach would be abandoned. The goal of sentencing would be to assess the degree to which the guilty citizen could be re-integrated into society. Eagleman has said, "To help a citizen re-integrate into society, the ethical goal is to change him *as little as possible* while bringing his behavior into line with society's needs." Eagleman acknowledges that not everyone who has violated social norms can be integrated into or back into (in the case of recidivism) society. Indeed, some people will have to be taken off the streets and incarcerated; some will have to be incarcerated for a lifetime because of their propensity toward recidivism.

Key questions that emerge from the determinism-free will debate are:

- Should every person found guilty of a crime be given a brain scan before sentencing to determine if there were some neurological anomalies that might have precipitated the criminal behavior?
- Before sentencing a person who has been found guilty of a crime should the guilty person's environmental history be reviewed before the jury to see how this history has interacted with the guilty person's neurological history?
- When someone's brain is scanned, be that person guilty of a crime, or be that person an upright citizen, who legally can have access to the results of the scan? This will become a very important issue for, as was shown above, more and more information (political preferences, purchasing preferences, sexual preferences, lying and truth telling, and so on) can be captured from a brain scan.
- Should brain scans be used for prevention purposes, and if the answer is "yes," to what extent should they be used? For example, if an individual exhibits behaviors that are suggestive of psychopathy, should that person be forced to undergo a brain scan, and if the results of that scan are predictive of psychopathic behavior, should that person be incarcerated?
- Is it ethical to scan someone's brain and, as a result of the scan, decide whether or not the person scanned is homicidal, racist, psychopathic, or does not have the quality of life that deserves sustenance?

## What is Death and Who Determines When Death Should Occur?

In 1990, at age twenty six, Terri Schiavo suffered a heart attack, and for ten to twelve minutes her brain received no oxygen. This resulted in brain damage. For about four years after Terri's trauma Terri's husband, Michael, and Terri's parents worked together on Terri's behalf. Then after a court case against Terri's fertility doctor, Michael Schiavo, received a settlement of one million dollars, three hundred thousand for loss of consortium and seven hundred thousand for Terri's care. At this point the relationship between Terri's husband and her parents turned into a feud that lasted until Terri's death. The autopsy showed that Terri Schiavo died of severe dehydration brought about because of the removal of her feeding/hydration tube. Terri died in 2005.

In the forward to the book titled, *The Case of Terri Schiavo* (Caplan, et al., 2006), Jay Wolfson wrote:

> Theresa Marie Schiavo may have had the most public death of any private person in history. Video snippets of her face and her partially clad, obviously disabled body were broadcast around the world on television and tabloids. Her name and her circumstances were front-page news for months. She captured the interest and attention of the Florida governor and catapulted its legislature into a special session. She was the singular cause for an extraordinary session of the U.S. Congress to be called during Easter week, and for the president of the United States of America to be transported by Air Force One from his Texas vacation to Washington, DC, to sign a law authored expressly for her. She activated all of the civil court systems at the state and federal levels, including the Supreme Court of the United States. She inspired Pope John Paul II to issue a confounding, nonencyclical statement admonishing against removal of artificial nutrition, and she stirred controversy within nearly all denominations about end of life decisions, medical technology, self-determination, and divine intent. (p. 13)

The feud between Terri Schiavo's husband and her parents intensified when in 1994 Terri's husband attempted to refuse treatment for an infection Terri suffered. Terri's family went to court and had that decision reversed. Thereafter the two camps fought an unending battle. Each action taken by the husband was reversed by court action or political persuasion initiated by the parents. This struggle was manifest by the battle of the feeding/hydration tube which was removed a total of three times, the third time by yet another court order which led to Terri's death

Terri's physicians declared that she was in a persistent vegetative state. Characteristic of this state is that the cerebral cortex is atrophied and is not functioning. The brain stem is usually functioning. Thus a patient in a persistent vegetative state can breathe without a respirator, can follow regular sleep and wakefulness cycles, can exhibit eye tracking, and can exhibit periods of alertness, for example. The diagnosis among neurologists, however, was not unanimous. Some neurologists said that she was not in this state or that they could not unequivocally declare that she was in this state. Writing in the *New England Journal of Medicine* in 2005, Thomas Quill said that the neurological tests indicated that she was in a persistent vegetative state. Quill also noted that it is common for family members to see signs of consciousness in persistent vegetative state patients. But these perceptions are not reality. If consciousness does not return within three or four weeks, it will almost never return, according to Quill. Quill, however, adds near the end of his essay that if the patient's wishes are not clear, then in the absence of pubic policy or family consensus, we should err on the side of continued treatment even in cases of persistent vegetative state in which there is no hope of recovery.

In the Terri Schiavo autopsy report (see Caplan et al. 2006), the chief medical examiner for Pasco and Pinellas counties in Florida, John Thogmartin, noted that Schiavo had an atrophied brain but ended his report in a waffling manner on the persistent vegetative state issue: "Neuropathologic examination alone of the decedent's brain—or any other brain, for that matter—cannot prove or disprove a diagnosis of persistent vegetative state or minimally conscious state."

The wrath against Terri's husband was intensified when he continually refused treatment for Terri after the first few years post trauma. For example, she lost five teeth because of a lack of dental care. There were marked restrictions placed on her and her surroundings. There was a restricted visitors' list, the shades were consistently pulled, and there were restrictions on the DVDs and CDs that she could watch or listen to, for example. In general, for the last sixteen years of Terri's life there was an impoverished environment with little or no stimulation that might serve to promote recovery.

One of the saddest aspects of the Terri Schiavo case is that she did not prepare a health care proxy and/or a living will. No one knew what Terri would have wanted. In most court situations hearsay evidence is not acceptable. But in the Schiavo case it was accepted because she could not speak for herself. A number of people noted that at the funerals of people who had been living with feeding/hydration tubes or other forms of mechanical life sustaining devices, Terri said that she would never want to live like that. This hearsay evidence is what convinced the court to order the removal of Terri's feeding/hydration tube for the last and final time, and the removal led to her death.

Quill (2005) claimed that when Terri was still living she was not consciously suffering and that dying in the way that she did can be a natural humane process: maybe yes, maybe no. Herzog (1999) has written an extensive case study of a locked-in brain patient, a young coed from Cornell University, who learned to communicate with trace eye movements that signaled "yes" and "no." Included in the case study were reports of traumatic events in the patient's life attributable to staff ignorance or insensitivity. For example, a podiatrist ripped off one of the patient's toenails without anesthesia. The pain was unbearable and the patient could say and do nothing about it. It is acknowledged, of course, that locked-in brain syndrome is different from Terri Schiavo's condition, but, at the same time, it is presumptuous to assume that Terri Schiavo felt no pain.

Questions about the diagnosis of persistent vegetative state have arisen since Schiavo's death. For example, Pinker (2007) described a case where a young woman was seriously injured in an automobile accident resulting in parts of her brain being crushed. The woman could open her eyes but she did not respond to sights, sounds, or jabs. The woman was judged to be in a persistent vegetative state. However, when an fMRI scan of the woman's brain was conducted it was found that when linguistic information was presented to the woman, language parts of the woman's brain lit up. Similarly, when the patient was asked to imagine moving in space or playing tennis, parts of the woman's brain responsible for these maneuvers lit up. As Pinker noted, this woman's brain scans were barely different from those of healthy volunteers.

Pinker asked his readers to try and comprehend what it would be like to be the woman described in the previous paragraph:

- Would you appreciate the words and caresses of your distraught family while racked with the frustration at your inability to reassure them that they are getting through?
- Or, might you drift in a haze, springing to life with a concrete thought when a voice prods you, only to slip back into blankness?
- And, if you could experience the existence suggested by one of the previous two questions, would you prefer that existence to death?

If the above questions have answers, Pinker mused, would the answers change our attitudes and our policies toward unresponsive patients? If our policies did change, the Terri Schiavo case would look like child's play, according to Pinker.

Monti et al. (2010) reignited the fires over life-sustaining care for those individuals thought to be in a persistent vegetative state. Their study used fMRI to scan the brains of fifty-four unresponsive patients thought to in a persistent vegetative state. Five of the patients showed clear fMRI responses to a variety of questions by indicating "yes" or "no." One patient had been unresponsive for five years. The study called into question the medical profession's methods for categorizing unresponsive patients. At best these methods seem to be very crude. It is worthwhile mentioning that there was no fMRI done on Schiavo because she had a wire implanted in her brain during an experimental procedure shortly after her trauma. The strong magnet in the fMRI scanner would have moved the wire causing more injury to her brain. Such fMRI data might have muted much of the debate about Schiavo which centered on her diagnosis as being in a vegetative state.

The *New England Journal of Medicine* published criticisms of the Monti et al. article on May 10, 2010. Following are two major criticisms of the article and the responses to these criticisms by two of the article's authors, Monti and Owen.

- *Criticism*: There is no evidence of willful modulation of brain activity in patients with non-traumatic brain injury (this would include Terri Schiavo). The authors responded that they had mentioned in their report that no fMRI responses were observed in any of the patients with non-traumatic brain injuries. The fear here of course is that many family members of patients judged to be in a persistent vegetative state from non-traumatic brain injury would have their hopes triggered that their loved one was really conscious.
- *Criticism*: There was no hierarchical testing of the patients. For example, do we know that the auditory functions of the brain were intact? The authors responded that if higher order functions were intact a good assumption would be that the lower functions in the hierarchy would be intact as

> well. Since the questions were asked orally in the Monti et al. study a good assumption would be that the auditory cortex was functioning as well. One might argue with the authors' response by saying that in the non-responsive patients it could be that dysfunctional lower level processes prevented any possible willful modulation attributable to higher order processes.

Monti and Owen end their reaction piece by noting that an fMRI scan is much less susceptible to bias than a bedside evaluation. They note, too, that little is known about the inner life of patients that have been diagnosed as being in a vegetative state. Finally, the authors argue for incorporating fMRI technology into the clinical process to address the misdiagnosis rate of 40 percent that has pervaded behavioral assessments.

Ethical questions that emerged from the Terri Schiavo case include:

- Should a spouse have the sole right to decide on his or her spouse's right to live or die if the suffering spouse has given no prior instructions in a living will or some other legal instrument?
- If the spouse should not have the sole right or if the brain injured person is not married, then who should make the decision? Should other family members be involved or should the sufferer's clergy be involved, for example?
- Is the use of a feeding/hydration tube considered to be a heroic measure?
- Should the person designated (by health care proxy or by court order) to make health care decisions for a brain injured individual have the right to deny all forms of health care including dental care?
- Is living in a persistent vegetative state dehumanizing?
- Is the treatment of brain injured individuals to be determined by the state or federal statutes?
- Is an insurance settlement or a legal judgment on behalf of a person judged to be in a persistent vegetative state to be paid to the person who made the decision to end the patient's life?

# Preemptive Actions

A key researcher and scholar in the area of the ethics of preemptive actions is Josh Greene (2003). Kristin Ohlson (2011) interviewed Greene, discussing his journey from philosophy to neuroscience. As a philosophy student, Greene was fascinated by two seemingly polar positions on ethics. First, was the almost absolutist view of Immanuel Kant. Kant believed that individuals have certain rights and that certain lines of behaviors are never to be crossed. On the other hand, is the utilitarian view of John Stuart Mill. Mill believed that the moral response was one that did the greatest good for the greatest number of people.

Greene experienced an epiphany when he read Antonio Damasio's *Descartes' Error*. Damasio concluded that reason and emotion were interrelated, on the basis of examining the case histories of individuals suffering brain damage. When emotional regions of the patients in these case histories were damaged, the patients could not make a decision.

Observers of modern warfare have noted that attackers are becoming more and more remote from their targets. Long-range artillery soldiers seldom see the devastation that they have caused. Bombing crews get only a glimpse of the destruction caused by their droppings. Fighter pilots flying at multi-mach speeds are essentially blind to the consequences of releasing their arsenals. Today, with drones, people sitting at computers in one country can be "piloting" drones (pilotless aircraft) on bombing runs over territory in another country halfway across the world. Less and less frequently do attackers see the consequences of their actions. It is interesting that a single photograph of a single person, particularly a child, can have a greater effect on an actor than all the destruction that has occurred at a remote distance from the actor who caused the destruction.

Of course, there are a number of ethical problems that exist or that can be created. Some of these have actually occurred. Five adults and a baby are hiding in a Dutch basement to escape Nazi capture. The Nazi soldiers are searching the house upstairs and the baby begins to cry. If the soldiers hear the baby crying five people and the baby will be captured and sent to a concentration camp. To stop the baby from crying requires that one of the adults cover the baby's mouth and nose and suffocate the baby. Should the baby be sacrificed to save five adults?

Another issue that Greene has considered is the human tendency to value life less as more lives are threatened. Stalin is reported to have said that a single death is a tragedy while the death of a million is a statistic. Why do we become riveted to the TV screen when a child is trapped in a well when at the same time we can become jaded by the deaths of thousands in a natural disaster? Greene takes an evolutionary approach and believes that our brains can only tolerate so much. He finds that when subjects are scanned while contemplating a number problem (saving one drowning man versus abandoning that victim to save a group of people whose boat has capsized), the insula and ventral striatum areas of the subjects' brains became active. These areas tend to manage probability and risk.

Greene takes an evolutionary approach to explaining the numbing effect that we experience as the number of victims in an event increases. He compares humans to squirrels. When a squirrel encounters a bountiful crop of nuts under a tree, the squirrel still takes only a certain number. First, there is the squirrel's cheek capacity; the squirrel can only pack so many nuts in its cheeks. But more importantly, perhaps, is the survival factor. A squirrel that spends too much time under the tree on any given occasion is exposed to danger. A predator may be watching from above or from the ground. The survival strategy for the squirrel is to get in, get the nuts, and get out. By analogy Greene believes that large numbers are essentially a threat. Just as a squirrel cannot stay too long under the tree in spite of there being a cornucopia of nuts there, so it is dangerous for a human to be concerned with large numbers of humans. Greene acknowledges that his evolutionary explanation is speculative but believes it has heuristic value, that is, it will promote further thinking on the topic.

Technology and global politics will continue to present troublesome ethical questions. Already there is a debate about using stem cells harvested from aborted fetuses to treat a number of pathologies. A potential recipient of stem cell treatment will probably not reject such treatment based on Greene's notion that one's aversion to a particular action is the degree to which one is physically involved in the action. Ordinarily a patient who is scheduled to receive stem cell treatment would have had no involvement in the abortion. In global politics politicians often talk about preventive strikes. Should a powerful nation bomb and destroy the nuclear facilities of another nation that appears to be manufacturing nuclear weapons to sell or to attack other nations. Here the numbers escalate. Is it ethical to kill one hundred thousand people to save the lives of one million people, for instance? Supporters of preemptive strikes often say that if the world had destroyed the Nazi military machine in the late 1930s, millions of lives could have been saved.

The importance of ethical issues in neuroscience is manifest in the organizations and programs that have formed to address the issues. For example, Oxford Centre for Neuroethics, established in 2009, describes their centre as follows:

> Neuroscience studies the brain and mind, and thereby some of the most profound aspects of human existence. In the last decade advances in imaging and manipulating the brain have raised ethical challenges, particularly about the moral limits of the use of such technology, leading to the new discipline of neuroethics. The Oxford Centre for Neuroethics, led by experts from ethics, philosophy of mind, neuroscience, neurology, psychiatry, and legal theory, will be the first international centre in the UK dedicated to neuroethical research. (see www. neuroethics.ox.ac.uk)

The International Neuroethics Society, headquartered in Bethesda, Maryland, USA, was formed in 2005. The Society describes itself as follows:

> We are an interdiscliplinary group of scholars, scientists, clinicians, and other professionals who share an interest in the social, legal, ethical, and policy implications of advances in neuroscience. The late 20th century saw unprecedented progress in the basic science of mind and in the treatment of psychiatric and neurologic disorders. Now, in the early 21st century neuroscience plays an expanding role in human life beyond the research lab and clinic. In classrooms, courtrooms, offices and homes around the world, neuroscience is giving us powerful new tools for achieving our goals and promoting a new understanding of ourselves as social, moral, and spiritual beings.

The society's mission statement is:

> Our mission is to promote the development and responsible application of neuroscience through interdisciplinary and international research, education, out-

> reach, and public engagement for the benefit of people of all nations, ethnicities, and cultures. (see www. neuroethicssociety.org)

An example of a program in neuroethics is Stanford's program in its School of Medicine. The program was initiated in 2005. It is described as follows:

> As a research team, we are devoted to the new field of neuroethics, with an initial focus on issues at the intersection of medical imaging and biomedical ethics. These include ethical, social, and legal challenges presented by advanced neurofunctional imaging capabilities, the emergence of cognitive enhancement neurotechnologies and pharmacology, self-referral to health care and imaging services, incidental findings, and fetal MRI. New initiatives are underway in regenerative medicine, neurogenetics and pediatric neuroethics.

The publisher, Springer, began publishing the journal, *Neuroethics,* with its first volume in 2008. The publisher has described the journal as a forum for interdisciplinary studies in neuroethics and related issues in the science of the mind. The professional societies and programs dedicated to the study of neuroethics and the journals that have been formed to serve as repositories to disseminate descriptions of research in the area serve as testimony to the importance of neuroscience and the issues raised by the endeavor.

## Summary

Did your brain make you do it? Are you just helplessly skidding along the greased axis of time? In addition, the establishment of guilt will become more difficult to establish. Equally difficult will be defining what is death and who makes the determination that death has occurred. Finally, if a brain scan reveals that a person will become a criminal, should some pre-emptive measures be taken to avoid that person's harming another. These are just some of the ethical questions that neuroscience will pose and society will have to solve.

# *Chapter 12*
# Future Trends

In Dante's *Inferno* we find that those who have predicted the future have been condemned to live forever in the inferno with their heads turned backwards. But this eternal threat is ignored by almost everyone. Parents make dire predictions to their children as to what will happen if they do not behave in a certain way, weather forecasters, day after day, make predictions about the weather, and scientists make predictions about physical and psychological phenomena. The purpose of this chapter, notwithstanding Dante's judgments of predictors, is to look at the future in neuroscience. What are some of the things that can be expected in the field? Acknowledged is the fact that many predictions will not come to pass. As mentioned in chapter 3 in this book, "Science proceeds one funeral at a time." Many predictions will be found not to be true, and many hypotheses will not be supported. Moreover, some long-held beliefs, assertions that have been supported by scientists for long periods of time, will be overthrown. The Arab proverb says, "The dogs may bark but the caravan moves along." And, as the actor/director Orson Wells said, quoting the Russian philosopher Ouspensky in his narration of the film, *Future Shock,* forty years ago: "Science is and should be unstoppable." In this chapter we will look at some predictions drawn from writings of scientists in neuroscience chemistry, in neuroscience technology, in pharmacology, in brain surgery, and predictions that will have immediate effects on the public at large.

## Future Trends in Neuroscience Biochemistry

Douglas Fields is a neuroscientist at the National Institutes of Health and an adjunct professor at the University of Maryland. His specialty is the glial cell.

For more than one hundred years glial cells have been considered little more than glue. But as Fields has noted (Fields 2011a and 2011b) glial cells are much more than that and, in point of fact, may replace neurons in importance since glial cells are the true workhorses of the brain. There are three basic kinds of glial cells: astrocytes, oligodendrocytes, and microglia. Astrocytes ferry nutrients and waste and mediate neuronal message transmission; oligodendrocytes insulate the axons with myelin, which facilitates the speedy transmission of neuronal signals; and microglial cells serve to fight infection and are called upon, when needed, to repair the brain. Without microglia the brain fails overall. Neurons make up 15 percent of the cells in the brain while glial cells make up 85 percent of the cells. Fields believes that glial research is bringing a new view of the brain which will change our perceptions of how the brain functions in health and disease. Echoing Kuhn, Fields believes that glial cell research is the beginning of a new research paradigm in neuroscience.

On the basis of his own research and the research of others, Fields believes that glial cells play a major role in both physical pathologies and psychopathologies. Included would be epilepsy, depression, schizophrenia, chronic pain, Alzheimer's, and the permanency of paralysis caused by spinal cord injury. Microglia are implicated in many of these diseases. These cells seek out and destroy attacking germs, clear away diseased tissue, and promote repair by releasing healing compounds. For example, in Alzheimer's disease amyloid plaques form in the brain and block the transmission of signals particularly between the hippocampus and the memory sites in the brain. Under normal circumstances, this plaque is digested by the microglia. But as people age, the microglia become weaker and atrophy, and the plaque is no longer destroyed and Alzheimer's "wins." Second, chronic pain seems to be mediated by neuron-glial cell interaction. In an effort to find a new and non-addictive pain killer, terminating this interaction is being considered. Third, there are a number of efforts underway to rehabilitate paralyzed spinal-cord-injured patients. Most basic among these efforts is to regenerate the cord at the site of injury and have it repair itself. One possibility here is to block the proteins that promote the sprouting of injured axons allowing damaged axons to re-grow. The analogy here would be that when a tree is cut sprouts will grow out of the stump but the felled tree will not reconnect with the stump. The goal here is to have the spinal cord reconnect. Last, the post-mortem analysis of brain tissue has also pointed to the possible role of glial cells in depression and schizophrenia. Manifested in these cases is a reduced number of glial cells. Fields has made note of the fact that almost all drugs used to treat mental illness and most neurological diseases work at the synapse controlling the release and reuptake of neurotransmitters. What regulates neurotransmitter levels? The glial cells called astrocytes.

Fields (2011a) certainly gives a Kuhnian-type statement when he says:

> Transformational moments are legendary in scientific history, but it is rare to witness one. Until quite recently, we neuroscientists had dismissed more than half of the brain as uninteresting—a humbling realization. We see only now

that the glial and neuronal brains work differently, and it is their intimate association that accounts for the astonishing abilities of the brain. Neurons are elegant cells, the brain's information specialists. But the workhorses? These are the glia.

## New Research on Neurons

While Fields sees the future in glial cells, Sanders (2011b) has reviewed a number of efforts to classify neurons on the basis of the neurons' shape and behavior. Previously neuroscientists classified neurons primarily on the basis of length, the shorter neurons serving as interneuron connections while those that descend into the spinal cord are quite long. Now additional dimensions are attracting the interest of neuroscientists. For example, classifying neurons on the basis of shape has led to theories about the information capacity of the brain. Some neuroscientists believe that a group of nerve cells, in which individual responses to electrical stimulation differ, is capable of processing more information than is a group of nerve cells where the responses are the same. The diversity of nerve cells may also have implications for treating a variety of brain disorders, dementia, schizophrenia, and autism, for example. Certain neurons may be associated with one of these diseases and not others. In addition, diversity in neurons may help an organism multi-task. For instance, a thief must focus on the task of stealing while at the same time be on the lookout for anyone who might witness the thievery. Finally, diversity in the nerve cells of the brain may allow natural selection to occur in the brain just as it does across species. Neurons that can adapt survive while those that do not adapt are reduced in number or become extinct.

## Brain Scans in the Future

At the age of ninety-three, Britton Chance, a professor emeritus of biophysics at the University of Pennsylvania, began developing a brain scanner that used beams of infrared light that passed harmlessly through the forehead and skull, penetrating the first few centimeters of cortical tissue (Silberman 2008). There the light bounced off the same concentrations of blood flow similar to the sensitivity of blood concentrations in traditional fMRI scans. When the light emerges from the cranium it is captured by optical sensors, then the "noise" is filtered out, and a scan is generated. Chance's prototype of the system used a Velcro headband, but the ultimate goal was to produce a device that would scan people's brains from across the room without the targets being aware that their brains were being scanned. The targets would not be in any scanner nor would they be wearing any gear. The issue here was if the device was successfully developed, could people have their brains scanned without their permission? It is a question of when individual liberties should be sacrificed for the safety of both self and others.

An article in the technology quarterly section of the September 3–9, 2011 issue of *The Economist* titled "Put your thinking cap on," reviews global research efforts dedicated to perfecting noninvasive BCIs using EEGs (p. 24 and 26). Progress is being made. Research teams in Austria, Spain, New Zealand, Britain, and the United States have made marked progress in the area. Research has focused on individuals with communication problems and paralysis. In addition, work has focused also on having normal individuals using thought to control many aspects of their daily lives such as controlling an automobile with thoughts alone. Marketers have not been bypassed as noted in the discussion of NeuroFocus in chapter 9. Of note is the progress in reducing the number of sensors needed to extract discrete messages from the brain. One manufacturer's headset looks much like a Bluetooth device, for instance. *The Economist* ended its report with a quote from Tan Le, the co-founder of Emotiv, an EEG headset maker: "We are only scratching the surface of what is possible," to which *The Economist* added, "Those scratches are getting deeper all the time."

## Individual Differences

If you chart almost any biological variable (height and weight, for example) by plotting the magnitude of the variable on the horizontal or x axis and frequency of occurrence on the vertical or y axis, you will get a distribution that looks very much like the normal curve. This is why statisticians can use the normal curve to make so many statistical inferences. These distributions also show that there are individual differences whenever we collect information about people. Not surprisingly, therefore, when we look at assessments of groups of individuals, we find differences. A future trend in neuroscience is to try to account for these individual differences. Several examples follow.

Stern (2006) has studied the confounding nature of these individual differences. In Alzheimer's, for instance, a common observation is that Alzheimer's patients show a buildup of amyloid plaque in the brain. The first question is, did the plaque cause the Alzheimer's or did the Alzheimer's cause the plaque, or did the two emerge together? The more fascinating question relates to individual differences. Stern found that post mortem examinations of Alzheimer's patients showed the presence of amyloid plaque in their brains. But, confounding the issue, was the fact that another group of individuals who, on post mortem examinations, showed plaque in their brains but who did not exhibit any Alzheimer's symptoms or demented speech when they were alive. What accounts for the differences?

Stern believes that the individuals who had the plaque but who did not exhibit Alzheimer's symptoms had, throughout their lifetimes, built up what he called cognitive reserve. A cognitive reserve buildup is the result of challenging your brain over a lifetime. The more cognitive reserve that you build up, the less you will be affected by brain disease as you age, it is believed. Cognitive reserve is thought to be manifested when alternative neural networks are established in the brain. Stern believes that these alternative networks are tied to the frontal lobes which are important in higher order brain functions. If the frontal lobes are not stimulated throughout life, the lobes will function less well in old age, according to Stern.

McGowan (2011) reported on the work of neurologists, Joseph Giacino and Nicholas Schiff with minimally conscious patients. One of their cases highlights the challenge and puzzlement of individual differences. A man fell into a minimally conscious state after a near drowning experience. For two years he did not speak, but he often groaned loudly at night. His mother, with whom he lived, purchased for her son a common sleeping pill to eliminate his nighttime groaning. After his first pill he seemed to become alert. The mother called his name, and he responded, "What?" Now after nine years the son can feed himself, answer questions, and show baseball grips as long as he takes his sleeping pills. The paradox is that the pill that helps many people sleep keeps this man awake.

## Future Trends in Surgery

When the drugs do not work or stop working the general course of treatment for a variety of serious diseases is invasive surgery and then less invasive surgery as advances are made in surgical procedures. For example, at one time the removal of the gall bladder (cholecystectomy) was a very invasive operation. Now that same surgery is much less invasive: it is done through laparoscopy. Always the hope is that the drugs can replace the surgery in treatment. This goal can be seen in the treatment for epilepsy. In the past, treatment for epilepsy commonly involved invasive brain surgery. Now such surgeries are done in only the most serious cases. Drug treatment has replaced surgery as a primary treatment.

Parkinson's disease has posed a particular problem. In Parkinson's the brain fails to produce a sufficient supply of the neurotransmitter, dopamine. The neurotransmitter, dopamine, is often called the reward neurotransmitter because it gives the rush that comes with addiction satisfaction whether it is alcohol, drugs, revenge, sex, winning, and so on. It is also responsible for movement and when the brain does not supply it, movement disorders emerge. A surgical procedure was developed that relieved some of the movement dysfunctions and speech disorders that characterize the disease. The surgery worked in some cases but not all. Then there was jubilation. A drug, Levodopa (L-dopa), had been developed that worked by replacing the dopamine that was no longer being produced by the brain. But jubilation was replaced by disappointment since, in some patients where the drug reduced or eliminated movement disorders and speech disorders, dyskinesic behaviors emerged. New solutions had to be sought.

Talan (2009) has presented a possible new solution designed primarily to treat Parkinson's disease but possibly other diseases as well. The possible solution that Talan discusses is deep brain stimulation. In this procedure, stimulating electrodes are strategically implanted in key areas of the brain. More specifically, two holes are drilled in the patient's skull. Through these holes are threaded wire leads with electrodes at the tip of the leads. Then the other ends of the leads are threaded under the skin to the upper chest where they are attached to a pacemaker-type device, called a neurostimulator, which has a power supply and is programmed to adjust the electrical signal to the brain in terms of frequency and intensity. After the implantation, a programmer begins to adjust the settings of the neurostimulator in order to find the optimum setting that leads to a reduction or elimination of symptoms. This programming can take months, and it is a clinical skill that some professionals have and others do not. There seems to be a very strong intuitive component to programming. The batteries of the system must be regularly checked and must be replaced every three to five years.

Talan recounts how this method has been used primarily to treat Parkinson's disease. Talan reports that of the approximately forty thousand people who have had the procedure, the vast majority have been Parkinson's patients. In about five to six percent of these patients the insertion site in the skull or the brain itself becomes infected as a result of the surgery. And, about twenty percent of the patients show post-implant involuntary dystonic movements that seem not to be related to Parkinson's disease. It has been found that deep brain stimulation is a technique that requires the mobilization of an entire team, neurosurgeons, neurologists, physiologists, psychiatrists, nurses, psychologists, speech and language therapists, physical therapists, occupational therapists, and ethicists, for example. Extensive efforts are being made for selecting patients for deep brain stimulation procedures. And as incidence data are collected, patients can be informed of their own singular risks. The risk of infection subsequent to implantation has been mentioned; there are also the general risks of surgery: the patient could develop bleeding and die. On the other hand, the patient may experience no symptom changes as a result of the surgery. Because of the severity of some symptoms experienced by many patients, the patients are willing to take the risks. A good predictive factor of a patient's experiencing success after deep brain stimulation is whether or not they experienced some success with L-dopa. If the taking of L-dopa produced symptom reduction for a period of time, then deep brain stimulation will have a better chance of working. Age is also a predictive factor. The older a patient is the less successful is deep brain stimulation. The same is true for patients with high blood pressure.

Parkinson's is a disease that severely affects both nonverbal and verbal communication. The nonverbal deficits are severe. With every movement or attempt at movement Parkinson's patients remind their communication partners of their movement pathologies. A comparison might be made with a person with a traumatic amputation of a limb. In a conversation the missing limb is noted but is soon forgotten. With severe movement disorders the conversational partner is continually reminded of the pathology. The same thing applies to speech. In the initial stages of Parkinson's, the speech is rarely affected. However, as the disease progresses voice problems often emerge and as the disease enters advanced stages, speech, language, and cognition are affected to the extent that, with every speaking turn by the Parkinson's patient, the communication partner is reminded once again of the patient's disease.

Besides Parkinson's disease, experimental trials with deep brain stimulation have been done on patients suffering from dystonia, obsessive-compulsive disorders, depression, Tourette's syndrome, epilepsy, pain, and minimal consciousness. In all these cases the application has to be characterized as experimental. About two hundred people have had deep brain stimulation for dystonia. In one study, twenty-two patients who had the procedure were followed for three years. The results of this study showed that the patients had improvements on their disabilities profiles of about 50 percent. Deep brain stimulation for both obsessive-compulsive and depressed patients seemed to work to some degree about 50 percent of the time. This may seem to be a rather anemic percentage but for patients suffering from these disorders the risk may be worth it, according to Talan. A small number of Tourette's syndrome patients have received deep brain stimulation for symptom relief. In many of these cases the results have shown marked symptom reduction or elimination. A marked reduction in the number of seizures has been found when epilepsy patients have been treated with deep brain stimulation. A finding of interest with epilepsy patients is something that may be labeled the *placebo effect*. After the electrodes are implanted and then turned off, patients will often continue showing a reduction in the number of their seizures. However, after a period of time the frequency of seizures may return to pre-surgical levels. Pain reduction using deep brain stimulation has had some success, often requiring repeated re-programming of the implant stimulator. Finally, a number of individuals have been awakened from their comas or persistent vegetative states using deep brain stimulation.

Some problems surface from these early studies with deep brain stimulation. Some suicides or attempted suicides have occurred with patients who have undergone the procedure. The percentages seem to fall in the one to two percent range. These data suggest that more efforts should be made to develop more stringent selection criteria and to inform patients comprehensively of the risks and rewards of the surgery. There are the perennial problems of pain and consciousness. One cannot feel another person's pain, and consciousness, as discussed earlier in this book, is a topic that many neuroscientists avoid. For minimally conscious patients or patients in a persistent vegetative state, a number of problems arise even with successful deep brain stimulation. The awakenings of these patients have sometimes created additional problems. With the return of partial or full consciousness but with almost total paralysis, patient care for extended periods of time is required. Who will pay for this care? And if the care is provided, who will be denied support because funds are being diverted to a newly awakened patient?

Deep brain stimulation is a cutting edge treatment for a number of disorders that affect communication between sufferers and their communication partners.

With Parkinson's patients the procedure seems to violate the usual hope in treatment, that is, drugs should replace surgery. And, in fact, when L-dopa became available, this hope was believed to have been realized. Now it is known that the positive effects of L-dopa may be limited. At the same time, when L-dopa does have a positive effect for a period of time, it can be a good predictor that deep brain stimulation will work when needed. One positive aspect of deep brain stimulation is that the stimulator can not only be programmed but it can be turned off completely. Deep brain stimulation is invasive, but it is less invasive than other surgical ventures. It is designed to deliver electrical stimulation to deep brain structures. In the future the basic procedure may become even less invasive with the electrical stimulation being delivered to the skull of the patient. This is a possibility that is probably years away, but it is a possibility. Of course, targeted drugs are also a possibility, and future trends in pharmacology will be discussed next.

## Future Trends in Pharmacology

Imagine that you can't find your car keys, then you can't find your cell phone, after which you can't find your car in a shopping center. Many people have experienced one or more of these memory lapses. However, if you do have your car keys, have found your cell phone and your car, but you can't find your way home then you probably have a serious memory problem. Hall (2007) has noted that in America there are probably four million people with Alzheimer's disease, twelve million more have mild cognitive impairments, and seventy-six million people are over fifty years of age. Among this large cohort of people, serious memory problems are emerging. Memory is intimately related to communication skills. It is impossible to tell another person of an experience unless you can remember the event. As the fastest growing segment of the population, the elderly pose a major problem for society in terms of memory loss. At the most extreme is that loss of memory found in Alzheimer's patients and the resultant dementia that accrues in later stages of Alzheimer's.

With this background in mind, it is not surprising that pharmaceutical companies are trying to develop "memory" pills. A number of drugs now on the market, although not all specifically memory drugs, have been prescribed (many times off label) to trigger brain responses compatible with better memory. Some of these drugs have been designed to treat mild to moderate Alzheimer's, and thus indirectly at least, are memory drugs. Others have been designed to treat narcolepsy or enhance attention. Narcolepsy drugs are often prescribed off label to help people stay awake, to sleep less, work harder, and stay focused during extended work periods. It is not surprising that the Defense Department is interested in narcolepsy drugs. On duty military personnel often need to stay awake and focused for extended periods of time. For example, on a particular aerial bombardment mission, pilots often need to stay awake and alert for extended periods of time. The drug Ritalin was designed to help children with Attention Deficit Hyperactivity Disorder (ADHD) focus but it is commonly used by others, particularly by some college students, to pull "all nighters" during exam periods. Acquisition of this prescription drug is relatively easy since it is so cheap and since it is often prescribed off label.

The future trend in drug development will involve not only the development of drugs to combat diseases but performance drugs as well. The latter category would include memory drugs and IQ drugs. People will wake up in the morning and decide if on this day they will need an IQ boost. To be sure, some of these drugs will have therapeutic value for people suffering from various diseases. In fact, some drugs will be designed for various pathologies but will be prescribed off label. It is quite possible that the so called memory drugs now in the pipeline at various drug companies will be used by people for whom the drug was not prescribed. Memory drugs can be designed to enhance memory from the number of digits a person can recall to more complex questions such as the names of the first forty presidents of the United States. This is rather straightforward. More elegant will be memory drugs that will inhibit frightening memories or facilitate the recall of pleasant memories. Such drugs might indeed help a person who suffers from panic attacks triggered by the recall of traumatic experiences. But those same drugs might be used by a person who simply wants to enjoy a day at home with pleasant memories.

## Future Trends in Neuromarketing

Plassmann et al. (2011) made a prediction about the future of neuromarketing which they refer to as consumer neuroscience. They note that presently consumer neuroscience involves large time, financial, and learning-curve costs. These costs would appear to place consumer research beyond the reach of many researchers. Plassmann et al. believe, however, that these barriers will wane in the near future and during the next decade there will emerge a flowering of interest in understanding the neural bases of consumer behavior.

## Other Future Trends

Restak (2006) and Lynch (2009) have provided their views of what will be happening in the future neurosociety. Restak believes that as citizens in the future neurosociety we will come to terms with:

- Tests aimed at revealing to others some of our thoughts and tendencies that, given our choice, we would choose to keep to ourselves.
- Brain scan images directed at gauging our suitability for certain jobs.

- Tests purporting to explain why we are romantically attracted to some people but repelled by others.
- Advertising campaigns that use brain scans to predict the products we are likely to purchase.
- Chemical enhancers of the brain designed to turn us into insatiable consumers, even for products and services we don't actually need.
- Brain image profiles of us to ascertain which political candidates we are most likely to vote for.
- Brain response patterns that reveal the emotions aroused in us by movies and television shows. (p. 2–3)

Lynch talks about the neurotechnological society. Characteristic of this society are the following:

- People will realize that they are not rational economic actors.
- Emergence of brain-computer interfaces that will expand an individual's capacity to parse data streams and accelerate profitable decision making.
- New forms of artistic expression will emerge. For example, virtual reality experiences will flourish.
- The relationships between the machine and the human mind will be elucidating.
- Destructive neuroweapons will be developed which will create a perpetual state of tension between promise and peril. (p. 16–17)

Restak's (2011) comments on the future of tracing neural connections in the human brain testify to the complexity of the organ referenced at the very beginning of this book. Restak has said quite simply that a map showing the neural connections in the human brain will not be achieved in the lifetime of anyone living today. We will, therefore, continue to rely on the functions of structures and regions in the study of the brain. Some of the key dynamics of the brain, particularly brain plasticity, will remain veiled.

## Summary

Overall, in terms of human communication, in the future more will be learned about what the brain is communicating, and more will be learned about communicating to the brain. Knowledge in both areas will be usable for good or evil. There will be a healthy symbiotic relationship between researchers working to solve the problems of individuals with brain-based communication problems and those working in mass communication areas like marketing and politics. The greatest test will be to have brains that can discriminate between good and evil and to have the wisdom to select the good.

# Bibliography


Allen, Paul. "Piece of Mind." *The Economist's The World in 2009* (2009): 157.

Andrews-Humma, Jessica, R. J. Reider, J. Spulcre, R. Poulin, and R.I. Buckner. "Functional-anatomic Fractionation of the Brain's Default Network." *Neuron*, 65, no.4 (February 25, 2010): 550–62.

Anthes, Emily. "Six Ways to Boost Brainpower." *Scientific American Mind* (February/March, 2009): 56–63.

Archer, Dane. *How to Expand Your S.I.Q. (Social Intelligence Quotient)*. New York: M. Evans, 1980.

Belmonte, Carlos. *The Dana Alliance's 2009 Progress Report on Brain Research: Introduction*. New York: Dana Press, 2009.

Bandettini, Peter. "Principles of Functional Magnetic Resonance Imaging." In *Functional Neuroimaging in Clinical Populations*. edited by Frank Hillary and John DeLuca. New York, NY: Guilford Press, 2007.

Bowlby, John. *Attachment: Attachment and Loss*. New York: Basic Books, Edition 1, 1969; Edition 2, 1999.

Bronowski, Jacob. *The Ascent of Man*. Boston: Little Brown & Co., 1973.

Brown, Alan. "Putting Thoughts into Action." *Scientific American Mind* (October/November 2008): 50–57.

Caplan, Arthur, James McCartney, and Dominic Sisti, eds. *The Case of Terri Schiavo: Ethics at the End of Life*. Amherst, NY: Prometheus Books, 2006.

Carlat, Daniel. "Mind Readers." *Wired*, June 2008: 119–128.

Carroll, John, ed. *Language Thought and Reality: Selected Writings of Benjamin Lee Whorf*. Cambridge, MA: The MIT Press, 1964.

Carter, Ruth, Susan Aldridge, and Martyn Page. *The Human Brain Book*. London: Dorling Kindersley, Ltd, 2009.

Chugani, Harry. Fine-Tuning the Baby Brain. *Cerebrum (A Dana Foundation Publication)*. July 1, 2004.

Claassen, D. O., K. A. Josephs, J. E. Ahlskog, M. B. Silber, M. Tippmann-Peikert, and B. F. Boeve. "REM Sleep Behavior Disorder Preceding Other Aspects of Synucleinopathies by Up to Half a Century." *Neurology*, 75 (2010): 494–99.

"Communicating Person to Person through the Power of Thought Alone." University of Southampton (October 6, 2009), http://www.soton.ac.uk/mediacentre/news/2009/oct/09_135.shtml (accessed September 7, 2011).
"The Complete Poems of Emily Dickinson." Boston: Little, Brown, and Company, 1929.
Condon, William and William Ogston. "Sound Film Analysis of Normal and Pathological Behavior Patterns." *The Journal of Nervous and Mental Disease,* 143, no. 4, 338–47, 1966.
"Correspondence." *New England Journal of Medicine* (May 20, 2010): 1936–38.
Cozolino, Louis. *The Neuroscience of Human Relationships*. London: W. W. Norton and Company, 2006.
Damasio, Antonio. *Descartes' Error: Emotion, Reason, and the Human Brain.* New York: HarperCollins, 1994.
Dance, Frank, and Carl Larson. *The Functions of Human Communication: A Theoretical Approach* (Appendix A). New York: Holt, Rinehart, and Winston, 1976.
Davis, Flora. *Inside Intuition: What We Know about Nonverbal Communication.* New York: McGraw-Hill, 1973.
Demos, Thomas. "Cirque du balancing act." *Fortune,* June 12, 2006, 114.
Dewey, John. *How We Think (1910).* Danvers, MA: General Books, LLC., 2009.
*Diagnostic and Statistical Manual of Mental Disorders: DSM-IV-TR*, Fourth Edition (Text Revision). Published by the American Psychiatric Association, 1994, Revised in 2000.
Durant, Will. *The Story of Philosophy*. New York: Pocket Books, 1974.
Eagleman, David. *Incognito: The Secret Lives of the Brain.* New York: Pantheon, 2011.
———. "The Brain on Trial." *The Atlantic*, July/August, 2011, 112–23.
Edelman, Gerald. *Second Nature*. New Haven, CT, 2006.
Egolf, Donald and Sondra Chester. *Forming Storming Norming Performing: Successful Communication in Groups and Teams*. New York: iUniverse, 2007a.
———. *The Nonverbal Factor*. New York: iUniverse, 2007b.
Ekman, Paul and Wallace Friesen. "The Repertoire of Nonverbal Behavior: Categories, Origins, Usage, and Coding." *Semiotica*, 69 (1969): 49–97.
Erikson, Erik. *Childhood and Society*. 2nd ed. New York: Norton, 1963.
Eyer, Diane. *Mother-Infant Bonding: A Scientific Fiction.* New Haven: Yale University Press, 1992.
Fields, Douglas. "The Hidden Brain." *Scientific American Mind* (May/June, 2011a): 53–59.
———. *The Other Brain.* New York: Simon and Schuster, 2011b.
Frank, Jerome. *Persuasion and Healing*. New York: Schocken Books, 1974.
Freud, Sigmund. *Psychopathology of Everyday Life*. London: General Book Publishers, 2009.
———. *The Interpretation of Dreams*. New York: HarperCollins (Avon Books), 1980.
Gao, W. et al. Evidence on the emergence of the brain's default network from 2-week-old to 2-year-old healthy pediatric subjects. *Proceedings of the National Academy of Sciences* 106, no. 16 (April 21, 2009): 679–685.
Gazzaniga, Michael. "Spheres of Influence." *Scientific American Mind*, (June/July 2008), 33.
———. *The Ethical Brain: The Science of Our Moral Dilemmas*. New York: Harper Perennial, 2005.
Gazzaniga, Michael, Richard Ivry, and George Magun. *Cognitive Neuroscience: The Biology of the Mind.* 3rd ed. New York: W.W Norton, 2009.
Goffman, Erving. *Behavior in Public Places*. New York: Free Press, 1963.

Goldberg, Jeffrey. "My Amygdala, My Self." *The Atlantic*, July/August, 2008, 95–100.

Goleman, Daniel. *Emotional Intelligence: 10th Anniversary Edition; Why it can Matter More Than IQ.* New York: Bantam Books, 2006.

———. *Social Intelligence: The New Science of Human Relationships*. New York: Bantam Books, 2007.

Granek, Jason et al. "The Effects of a Video Game Experience on the Cortical Networks for Increasingly Complex Vasomotor Tasks." Paper presented at Society for Neuroscience Conference Program. August 24, 2007.

Green, Shawn and Daphene Bavelier. "The Cognitive Neuroscience of Video Games." In *Digital Media: Transformations in Human Communication,* eds. P. Messaris and L. Humphries. New York: Peter Lang.

Greenberg, Gary. "The Book of Woe: Inside the Struggle to Define Mental Illness." *Wired*, January, 2011, 124–36.

Greene, Joshua, Brian Sommerville, Leigh Nystrom, John Darley, and Jonathan Cohen. "An fMRI Investigation of Emotional Engagement in Moral Judgement." *Science,* 293, (2003): 2105–2108.

Grim, Patrick. *Philosophy of Mind: Brains, Consciousness, and Thinking Machines.* Chantilly, VA: The Teaching Company, 2008.

Guenther, Frank. "Cortical Interactions Underlying the Production of Speech Sounds." *Journal of Communication Disorders*, 39, no. 5, (September/October 2006): 350–65.

Hall, Stephen. "The Quest for a Smart Pill." In Floyd Bloom, ed. *Best of the Brain from Scientific American.* Washington, D.C: Dana Press, 2007, 233–46.

Harper, Nancy. *Human Communication Theory: The History of a Paradigm.* Rochelle Park, NJ: Hayden Book, 1979.

Helman, Christopher. "Inside Boone's Brain." *Forbes* (June 30, 2008): 76–78.

Herzog, B. "Locked In." *Cornell Magazine Online,* July/August, 1999.

Hess, Eckerd, "Attitude and Pupil Size." *Scientific American*, 212, (1965): 46–54.

——— "The Role of Pupil Size in Communication." *Scientific American*, November, 1975.

Hilgard, Ernest R. *Theories of Learning.* 2nd ed. New York: Appleton-Century-Crofts, 1956.

Hochberg, Leigh. "Turning Thought into Action." *New England Journal of Medicine,* 359, no.11, (2008): 1175–77.

Hochberg, Leigh, Arto Nurmikko, and John Donoghue. "Brain Machine Interface." *Annual Review of Biomedical Engineering* 13 (August 2011).

Hochberg, Leigh, M. D. Serruya, G. M. Friehs, J. A. Mukand, M. Salch, A. H. Caplan, A. Branner, D. Chen, R. D. Penn, and J. P. Donoghue. "Neuronal Ensemble Control of Prosthetic Devices by a Human with Tetraplegia." *Nature* 442, no. 7099 (July 13, 2006): 164–71.

Iacoboni, Marco. *Mirroring People*. New York: Farrar, Straus, & Giroux, 2008.

Ibbotson, Jimmy. The quote is transcribed from an interview with Ibbotson on a video recording of Bill Gaither Presents "A Gospel Bluegrass Homecoming," Volume 1, Alexandria, IN: Gaither Music Group: 2003.

Ioannidis, John. "Why Most Published Research Findings Are False." *PLoS Medicine,* 2, no. 8 (August 30, 2005):e124.

James, William. *Psychology: The Briefer Course*. Mineola, NY: Dover Publications, 2001. This book was originally published by Henry Holt in 1892.

Jones, Edward and Lorne Mendell. "Assessing the Decade of the Brain." *Science* (April 30, 1999): 739.

Jung, Carl. *Memories, Dreams, Reflections.* New York: Random House, 1965.

Karen, Robert. *Becoming Attached: First Relationships and How They Shape Our Capacity to Love.* New York: Oxford University Press, 1998.

Kennedy, Philip and Robert Bakay. "Restoration of Neural Output from a Paralyzed Patient by a Direct Brain Connection." *NeuroReport* 9, no. 8 (June 1, 1998): 1707–11.

Kiefer, Ingrid. "Brain Food." *Scientific American Mind* (October/November 2007): 58–63.

Klaus, Marshall and John Kennell. "Maternal Attachment: Importance of the First Postpartum Days." *New England Journal of Medicine* 286, no.9 (March 2, 1972): 460–63.

Knutson, Brian, Scott Rick, Eliot Wimmer, Drazen Prelec, and George Lowenstein. "Neural Predictions of Purchases." *Neuron* 53, no. 1 (January 4, 2007): 147–56.

Kuhn, Thomas. *The Structure of Scientific Revolutions.* Chicago, IL: University of Chicago Press, 1962.

Lakoff, George. *The Political Mind.* New York: Penguin Group, 2008.

Lamb, M. "Second Thoughts on First Touch." *Psychology Today,* April 1982, 9–11.

Larson, Christine. "Keeping Your Brain Fit." *U.S. News and World Report*, February 11, 2008, 41–49.

Lehrer, Jonah. "The Truth Wears Off: Is There Something Wrong with the Scientific Method?" *The New Yorker*, December 13, 2010, 52–57.

Lehrer, Joseph. *How We Decide.* New York: Houghton Mifflin Harcourt, 2009.

Leiber, Nick. "Speaking Simply." *Bloom Business Week,* August 15–August 28, 2011, 40.

Levine, Amir and Rachel Heller. *Attached: The New Science of Adult Attachment and How It Can Help You Find—And Keep—Love.* New York: J. P. Tarcher of Penguin Group, 2010.

Levitin, Daniel. *This is Your Brain on Music.* London: Penguin Books Ltd., 2006.

Lindstrom, Martin. *buy.ology.* New York: Doubleday/Random House, 2008.

Littlejohn, Stephen. *Theories of Human Communication.* Columbus, OH: Charles E. Merrill, 1978.

Lynch, Zack. *The Neuro Revolution: How Brain Science Is Changing Our World.* New York: St. Martin's Press, 2009.

MacDonald, Matthew. *Your Brain: The Missing Manual.* Sebastopolo, CA: Pogue Press/O'Reilly Media, 2008.

Marcus, Gary. *Kluge: The Haphazard Construction of the Human Mind.* New York: Houghton-Mifflin, 2008.

McClure, Samuel, Li Jian, Damon Tomlin, Kim Cypert, Latane Montague, Read Montague. "Neural Correlates of Behavioral Preference for Culturally Familiar Drinks." *Neuron* 44, no. 2 (October 14, 2004): 379–87.

McGowan, Kat. "Back From the Brink." *Discover*, March 2011, 62–71, 76.

Medina, John. *Brain Rules.* Seattle, WA: Pear Press, 2008.

Merker, Bjorn. "Consciousness without a Cerebral Cortex: A Challenge for Neuroscience and Medicine." *Behavior and Brain Science* 30, no. 1 (2007): 63–81.

Mildner, Vesna. *The Cognitive Neuroscience of Human Communication.* New York: Lawrence Erlbaum (Taylor & Francis Group), 2008.

Miller, Johnathan. *The Body in Question.* New York: Random House, 1978.

Monti, Martin, Audrey Vanhaudenhuyse, Martin Coleman, Melanie Boly, John Pickard, Luaba Tshibanda, Adrian Owen, and Steven Laureys. "Willful Modulation of Brain Activity in Disorders of Consciousness." *New Engl Journal of Medicine* 362 (February 18, 2010): 579–89.

Mossman, Kaspar. "Brain Trainers." *Scientific American Mind*, (April/May, 2009): 32–39.

Murray, David. *Borrowing Brilliance*. New York: Gotham Books, 2009.

Nicolelis, Miguel. *Beyond Boundaries: The New Neuroscience of Connecting Brains with Machines—and How It Will Change Our Lives*. Times Books, 2011.

Noordzij, Matthijs, Sarah Newman-Norlund, Jan de Ruiter, Peter Hagoort, Stephen Levinson, and Ivan Toni. "Brain mechanisms underlying human communication." *Frontiers in Human Neuroscience* 3, no. 14 (2009): 1–13.

Norden, Jeanette. *Understanding the Brain*. Chantilly, VA.: The Teaching Company, 2007.

Ogawa, S., T. M. Lee, A. R. Kay, & D. W. Tank. "Brain magnetic resonance imaging with contrast dependent on blood oxygenation." *Proceedings of the National Academy of Sciences, USA*, 87 (1990): 9868–9872.

Ohlson, Kristin. "The End of Morality." *Discover*, July/August, 2011, 32–36; 92.

Park, Alice. "Family History: High Risk of Autism among Siblings." *Time*, August 29, 2011, 19.

Parker, I. "Reading Minds," *The New Yorker*, January 20, 2003, 52–63.

Penenberg, Adam. "They Have Hacked Your Brain." *Fast Company*, September, 2011, 85–9, 123–25.

Penfield, Wilder. "The Recording of Consciousness and the Function of Interpretive Cortex." Chapter III in Penfield, Wilder and Lamar Roberts. *Speech and Brain-Mechanisms*. New York: Atheneum, 1966.

———. *The Mystery of the Mind*. Princeton, NJ: Princeton University Press, 1975.

Penfield, Wilder and Lamar Roberts. *Speech and Brain-Mechanisms*. New York: Atheneum, 1966.

Perls, Frederick. "Four Lectures" In Fagan, et al., eds. *Gestalt Therapy Now*. Palo Alto, CA: Science and Behavior Books, 1970.

Pfaff, Donald. *The Neuroscience of Fair Play*. New York: Dana Press, 2007.

Pinker, Steven. "The Brain: The Mystery of Consciousness." http://www.time.com/ magazine/article/0,9171,1580394,00.html, January 10, 2007. Viewed May 17, 2011.

Plassmann, Hilke, Carolyn Yoon, Fred Feinberg, and Baba Shiv (2011). Consumer Neuroscience, edited by R. P. Bagozzi and A. A. Ruvio. *Wiley International Encyclopedia of Marketing* 3, 115–22. West Sussex, U.K.: John Wiley & Sons.

Plassmann, Hilke, John O'Doherty, Baba Shiv. and Antonio Rangel. "Marketing Actions Can Modulate Neural Representations of Experienced Pleasantness." *Proceedings of the National Academy of Sciences*, No. 105, 2007, 1050–54.

Pollack, M. "Intelligent Technology for an Aging Population: The Use of AI to Assist the Elderly with Cognitive Impairment." *AI Magazine*, Summer 2005.

Purves, Dale. *Brains: How They Seem To Work*. Upper Saddle River, NJ: Pearson Education, 2010.

Quill, Timothy. "Terri Schiavo—A Tragedy Compounded." *New England Journal of Medicine* 352, no. 16 (2005): 1630–33.

Raichle. Marcus. "The Brain's Dark Energy," *ScientificAmerican* (March 2010): 44–49.

Raichle, Marcus, Ann Mary MacLeod, Abraham Z. Synder, William J. Powers, Debra A. Gosnard, and G. I. Shulman. "A Default Mode of Brain Function," *Proceedings of the National Academy of Sciences of the United States of America* 98, no. 2 (January 16, 2001): 676–82.

Restak, Richard. *Think Smart*. New York: Riverhead Books, 2009.

———. *Optimizing Brain Fitness*. Chantilly, VA: The Teaching Company, 2011.

——— *The Naked Brain: How the Emerging Neurosociety is Changing How We Live, Work, and Love*. New York, NY: Harmony Books, 2006.

Roberts, Royston M. *Serendipity: Accidental Discoveries in Science*, New York: John Wiley and Sons, 1989.

Robinson, Daniel. *Consciousness and Its Implications*. Chantilly, VA: The Teaching Company, 2007.

Roy, C.S. and Sherrington, C.S. "On the Regulation of the Blood-Supply of the Brain." *Journal of Physiology*, 11(1–2) (January, 1890): 85–158.

Saenz, Aaron. "Another Attempt to Use fMRI Lie Detection in U.S. Court Fails in Brooklyn. More on the Way." Filed on May 6, 2010 and printed on the Singularity Hub, July 23, 2011.

Saey, Tina Hesman. "You Are Who You Are by DEFAULT." *Science News*, (July 18, 2009): 16–20.

———. "Let There Be Light." *Science News*, (January 30, 2010): 18–21.

Samson, Kurt. "Will the Demise of the NIH Human Brain Project Hurt the Growth of Neuroinformatics?" *Neurology Today* 6, no. 5 (March 2006): 34–35.

Sanders, Laura. "Residents of the Brain." *Science News*, July 20, 2011b, 22–25.

———. "One Problem, Many Paths: Autism's Many Genetic Players May Act through Common Networks." *Science News*, August 13, 2011a, 20–21.

Sapolsky, Robert. *Biology and Human Behavior: The Neurological Origins of Individuality* 2nd ed. Chantilly, VA: The Teaching Company, 2005.

Saporta, Sol. *Psycholinguistics: A Book of Readings*. New York: Holt, Rinehart and Winston, 1961.

Seabrook, John. "Suffering Souls." *The New Yorker*, November 10, 2008, 64–73.

Shermer, Michael. "Why You Should Be Skeptical of Brain Scans." *Scientific American Mind*, October/November, 2008, 67–71.

Silberman, Steve. "The Cortex Cop." *Wired*, January 2008: 146.

Simeral, John, Sun-Phil Kim, Michael Black, John Donoghue, and Leigh Hochberg. "Neural Control of Cursor Trajectory and Clicking By a Human with Tetraplegia 1000 Days After Implant of an Intracortical Microelectrode Array." *Journal of Neural Engineerin*, 8, no. 2 (April, 2011).

Sperry, Roger. "Left Brain Right Brain." *Saturday Review*, August 9, 1975, 30–33.

Stern, Y. *Cognitive Reserve: Theory and Application*. London: Psychology Press, 2006.

Sternberg, Eliezer. *My Brain MADE ME DO IT*. Amherst, NY: Prometheus Books, 2010.

Talan, Jamie. *Deep Brain Stimulation*. Washington D.C: The Dana Press, 2009.

Talbot, Margaret. "Duped: Can Brain Scans Uncover Lies?" *The New Yorker*, July 2, 2007, 52–61.

Tauber, Robert. *Self-fulfilling Prophecy*. Westport, CT: Praeger Publishers, 1997.

Toobin, Jeffrey. "The Mitigator." *The New Yorker*, May 9, 2011, 32–39.

Touhami, Zineb, Larbi Benlafkih, Mohamed Jiddane, Yahya Cherrah, Hadj Malki, and Ali Benomar. "Neuromarketing: Where Marketing and Neuroscience Meet." *African Journal of Business Management* 5, no. 5 (March 4, 2011): 1528–32.

University of Southampton. "Brain Computer Interface Allows Person-to Person Communication through Power of Thought." *ScienceDaily*. Retrieved October 8, 2009, from http://www.sciencedaily.com/releases/2009/10/09/1006102637.htm

Valenstein, Eliot S. "Otto Loewi: An Autobiographical Sketch," *Perspectives in Biology and Medicine*, Autumn, 1960.

Van De Castle, Robert L. *Our Dreaming Mind*. New York: Ballantine Books/Random House, 1994.

von Bekesy, Georg. *Experiments in Hearing*. New York: McGraw-Hill, 1960.

Vul, Edward, Christine Harris, Piotr Winkielman, and Harold Pashler. Puzzlingly High Correlations in fMRI Studies of Emotion, Personality, and Social Cognition. *Perspectives in Psychological Sciences* 4, no. 3, (2009): 274–90.

Vygotsky, Lev. "Thought and Speech," edited by Sol Saporta, Sol. *Psycholinguistics: A Book of Readings*. New York: Holt, Rinehart and Winston, 1961.

Warner, Judith. "Autism's Lone Wolf." *Time*, August 29, 2011, 44–7.

Westen, Drew. *The Political Brain.* New York: Perseus Books Group (Public Affairs), 2007.

Wolpaw, Jonathan, Niels Birbaumer, Dennis McFarland, Gert Pfurtscheller, and Vaughan. "Brain-computer Interfaces for Communication and Control." *Journal of Clinical Neurophysiology* 113 (2002): 767–91.

Zweig, Jason. *Your Money & Your Brain.* New York: Simon & Schuster, 2007.

# Index

# *AUTHOR NOTES*

**Donald B. Egolf** is an Associate Professor of Communication at the University of Pittsburgh where he teaches undergraduate courses in Interpersonal, Small Group, and Nonverbal Communication, and graduate courses in Communication Research. Egolf's interest in the brain began in a graduate physiological psychology class when the instructor demonstrated the effects of electrical stimulation to the reward center of the brain of a laboratory rat in a Skinner box. This interest increased in both breadth and depth with Egolf's early career as a certified speech and language pathologist working with brain injured patients, primarily stroke patients. One non-stroke patient that sparked Egolf's interest even further was a twelve-year-old boy who had a hemispherectomy (the left side of the boy's brain was removed to relieve symptoms of severe epilepsy at age eight). Egolf worked with the boy for a year in an effort to help him regain speech and language functions. As the years went by Egolf's interests in the brain and communication expanded to wider contexts, particularly in the interpersonal and mass media contexts. The discovery of mirror neurons suggested a neural basis for a number of interpersonal phenomena, and the Coke versus Pepsi studies as well as the wine preference studies began to establish a neural foundation for many of the present day mass-audience persuasive concepts and strategies whose roots go back as far as ancient Greece. Egolf's interest in the brain reflects the recent interest in this area by a number of scholars in a variety of disciplines.